DEVELOPMENTAL AND CELL BIOLOGY SERIES

EDITORS
M. ABERCROMBIE   D. R. NEWTH
J. G. TORREY

MAMMALIAN CHIMAERAS

MAMMALIAN CHIMAERAS

ANNE McLAREN, FRS

Director, Medical Research Council Mammalian Development Unit

CAMBRIDGE UNIVERSITY PRESS

CAMBRIDGE

LONDON · NEW YORK · MELBOURNE

Published by the Syndics of the Cambridge University Press
The Pitt Building, Trumpington Street, Cambridge CB2 1RP
Bentley House, 200 Euston Road, London NW1 2DB
32 East 57th Street, New York, NY 10022, USA
296 Beaconsfield Parade, Middle Park, Melbourne 3206, Australia

First published 1976

Printed in Great Britain
at the
University Printing House, Cambridge
(Harry Myers, University Printer)

Library of Congress Cataloguing in Publication Data

McLaren, Anne.
Mammalian chimaeras.

(Development and cell biology series; no. 4)
Bibliography: p.
Includes index.
1. Mosaicism. 2. Mammals–Genetics. I. Title.

QH445.7.M32   599'.03   75-40988

ISBN 0 521 21183 2

Contents

Preface

This book is on a very specialized topic. The few dozen people in the world who have worked with experimental chimaeras will share my enthusiasm for their beauty, their unexpectedness, the insight that they provide into old questions, and above all for the new questions that they continually raise, questions that one never dreamt existed in the days when an individual had two parents only. If this book communicates some part of this fascination to a few other people interested in mammalian development (and who isn't?), I shall be satisfied.

My aim in the bibliography has been to include all papers published on primary chimaeras in mammals up to the end of 1974, thus enabling myself to throw out large numbers of reference cards. If I have missed any papers, my apologies to their authors.

For the time and effort involved in reading chapters of the book in draft, and for their valuable comments, I am very grateful to Professor Donald Michie, Dr Richard Gardner, Dr Chris Graham, Dr Charles Ford, FRS, Dr Bruce Cattanach, Dr John West, Dr Katie Bechtol, Dr Ron Barnes and Professor David Newth. I would also like to thank Hilary Prout for her meticulous and enthusiastic typing and reference-chasing.

September 1975   ANNE MCLAREN

1
What is a chimaera?

A chimaera is a composite animal. The Chimaera of Homeric legend had the body of a she-goat, the head of a lion, and the tail of a serpent. Many other combinations are described in the literature of antiquity: the six-limbed centaur, half man, half horse; the harpy, a bird of prey with the head and breasts of a woman; the griffon, with eagle's head and legs, and the body of a lion; the beast of the Apocalypse, like a seven-headed leopard with the mouths of a lion and the feet of a bear; the Apocalyptic locusts, horse-shaped with men's faces, women's hair, lions' teeth and scorpions' tails. All bear witness to the ambition of Man to combine the outstanding qualities of different animals into a single creature of surpassing power. Sexual hybridization is at best a chancy method for this purpose: it may take several generations, the results are unpredictable and it can only be used within a species.

Biology as well as mythology provides examples of strange and often intimate associations between different species. Alga and fungus form a partnership so close that we refer to it by a single name, 'lichen'; the hermit crab collects stinging sea anemones to guard its shell; the sea slug (*Aeolis pilata*) accumulates nematocysts from the hydroid that it eats, and positions them in its epidermis as a defence. Even some subcellular organelles, like chloroplasts and mitochondria, are now thought to have originated as independent organisms existing in symbiotic relationship to the host cell.

Some definitions

Today, 'chimaera' is used to describe any composite animal or plant in which the different cell populations are derived from more than one fertilized egg, or the union of more than two gametes. Such a situation can arise in various ways. For convenience, Ford (1969) has distinguished between 'secondary chimaerism', where tissues are combined from two or more adult individuals, or from embryos after the period of organogenesis has begun, and 'primary chimaerism', where the genetically different cell populations co-exist from a very early stage of embryogenesis, even from fertilization.

Secondary chimaeras can be formed by grafting or transplantation of tissues, for example kidney grafting, skin grafting or blood transfusion, or can arise spontaneously, by transfer of cells in mammals between mother

Fig. 1. Etruscan chimaera, made of bronze, combining structures of lion, serpent and goat. From *Cambridge Ancient History*, Vol. 1 of Plates, pp. 336–7. London, Cambridge University Press. Reproduced by kind permission of the Mansell Collection.

and embryo, or from one embryo to another. For example, a pair of twins studied by Booth *et al.* (1957) both had a mixture of red cells in their blood: Mr Wa. had 86 % group A and 14 % group O cells, while his twin sister Miss Wa. had 99 % O and 1 % A cells. The two cell populations proved to differ also at the *Rhesus, Duffy, Kidd* and *Dombrock* loci. Such individuals raise critical immunological problems. An analogous situation can be produced experimentally by giving an animal a lethal dose of radiation to knock out its own immune system, and then 'rescuing' it by a bone marrow transplant from a genetically different individual. The study of 'radiation chimaeras' has contributed greatly to transplantation biology (Micklem & Loutit, 1966).

In cattle, twin embryos almost always share a common placental circulation, and secondary chimaerism can readily be detected in the blood. Unlike the situation of Mr Wa. and his sister, if one component predominates in the blood of one member of a cattle twin pair, the same component tends to be in the majority in the other twin also, at least in the adult (Marcum, 1974). When the twins are of opposite sexes, the female nearly always develops as a sterile freemartin (a phenomenon known already in Roman times), and the male co-twin is also affected. The impact of secondary chimaerism on sexual development is discussed briefly in Chapter 4. Sheep, goat and pig twins only occasionally show blood chimaerism and freemartinism, because placental vascular anastomoses seldom occur, but the rare twin pairs studied in horses show no signs of freemartinism even when vascular anastomoses and blood chimaerism are present. Marmosets produce twins in almost every pregnancy, and blood chimaerism is the rule, but they show no abnormalities of the reproductive system.

Primary chimaeras can be formed by aggregating or combining cells from different embryos at a very early stage of development, or can arise spontaneously at fertilization if more than one spermatozoon fertilizes a single egg, or an egg and its polar body. 'Fertilization chimaeras' will be considered further in Chapter 11, in the context of spontaneous chimaerism in Man and other animals. The first category of primary chimaerism, in which cells from different embryos are brought together experimentally to form a single individual, constitutes the subject matter of most of the rest of this book.

A different term, 'mosaic', is used to describe a composite individual derived from a single fertilized egg, that is from the union of one egg with one spermatozoon. Benirschke (1970) has argued that there is no difference in principle between chimaeras and mosaics, and certainly it is often difficult in spontaneously composite individuals to ascertain how the condition arose. In practice, however, the distinction can be useful.

In a mosaic individual, the different cell populations must arise during the course of development, by some process such as somatic mutation, somatic recombination or non-disjunction, spontaneous or induced. For example, a mosaic with an XX (or XY) and an XO cell line would be formed if an XX (or XY) zygote lost a sex chromosome at an early cleavage division, giving rise to an XO cell line as well as the original XX (or XY) line. On the other hand the occasional individual reported to have both an XX and XY cell line is unlikely to be a mosaic, since such a situation cannot arise by any simple non-disjunctional event: it could however be a chimaera, formed by aggregation of two separate zygotes, or by double fertilization, with two separate spermatozoa fertilizing an egg and its polar body. To take another example, female mammals (eutherians anyway) are mosaic for any X-linked locus for which they happen to be heterozygous, since half of their cells will have one X chromosome inactivated and half the other. For the same X-linked locus, a chimaera could also in principle be constructed, by aggregating embryos from two strains homozygous for different alleles at that locus. Both the mosaic and the chimaera would have two cell populations, with alternative alleles active; the mosaicism would have been initiated whenever X chromosome inactivation occurs, probably soon after implantation, while the chimaerism would have begun earlier, as embryo aggregation is usually carried out at the 8-cell stage.

The similarities and differences between mosaics and experimental chimaeras in mammals will be discussed further in Chapter 9.

Experimental chimaeras in animals

Experiments on lower vertebrates, in which embryos or parts of embryos are aggregated together, have mainly used later embryos, forming 'secondary chimaeras' in the sense defined earlier. The early work by Spemann and his colleagues on determination of the neural plate (see Spemann, 1938) involved grafts at the gastrula stage between *Triturus* embryos differing in degree of pigmentation, while in later induction experiments composite embryos were formed at the gastrula or early neurula stage, with ectoderm and mesoderm from one individual and endoderm from another (e.g. Okada, 1955). Studies on the origin of primordial germ cells in Amphibia have involved grafts at the neurula stage, either of the actual area containing primordial germ cells in *Xenopus* (Blackler & Fischberg, 1961; Blackler, 1962) or of the entire endoderm or mesoderm in urodeles (Nieuwkoop, 1946). The transplantations were made between different subspecies or species differing in egg size and colour, and the chimaeric individuals were reared to the adult stage so that the developmental origin of their eggs could be determined.

Attempts to produce primary chimaeras in invertebrates and lower vertebrates have rarely been successful. Early experiments on sea urchins and Amphibia have been described by Morgan (1927). George (1958) treated the egg capsules of snails with sodium tauroglycocholate in order to weaken the vitelline membrane and thus facilitate aggregation of the eggs, and obtained two conjoined twin monsters, showing head-to-head union, out of 200 treated capsules (Fig. 2). Presumably the determinate mode of development characteristic of most invertebrate species does not permit the full regulation of two embryos into a single normal individual, as can occur in mammals. On the other hand, some insect gynandromorphs, with two cell lines of different sex, are thought to be chimaeras formed as a result of double fertilization (for review, see Stern, 1968). Recently, chimaeric *Drosophila* have been produced by transplanting nuclei into specific regions of the fertilized egg (Okada, Kleinman & Schneiderman, 1974).

In birds, secondary chimaeras have been produced by joining the blood circulations of two fertilized eggs, a procedure known as 'parabiosis'. After hatching, the chicks can both be shown to have not only two populations of blood cells, but also two populations of germ cells, since in birds the primordial germ cells circulate in the embryonic blood. Primary chimaeras have been produced by Marzullo (1970) by injecting chick blastodisc cells from pigmented Barred Rock or Rhode Island Red eggs into unincubated host eggs of the unpigmented White Leghorn breed. Of 239 eggs injected, nineteen survived to the stage of feather formation, but only three showed evidence of pigmentation, indicating chimaerism (Fig. 3). In each case the pigment was of donor type.

The systematic study of primary chimaeras in mammals began in 1961,

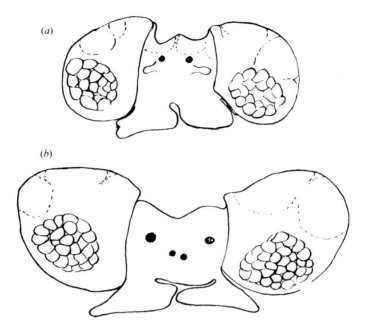

Fig. 2. Twin monsters, (*a*) 7 days old and (*b*) 10 days old, showing head-to-head union of embryos of the snail *Limnaea stagnalis*, facilitated by treatment with sodium tauroglycocholate. From George (1958).

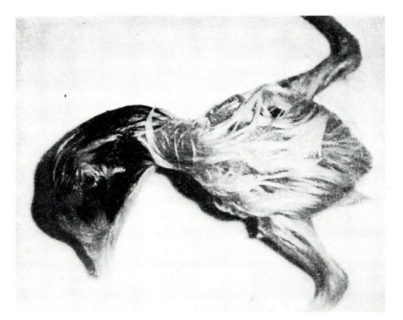

Fig. 3. Chick chimaera, aged 15 days, composed in part of White Leghorn cells and in part of Barred Rock cells. From Marzullo (1970).

when Tarkowski reported the successful aggregation of cleaving mouse embryos. Embryos aggregated in pairs formed single large blastocysts in culture. After transplantation to the uterus of a foster-mother, these blastocysts developed into foetuses of apparently normal size and morphology, though the sex ratio was abnormal and one hermaphrodite was observed (see Chapter 4). Some of the embryos were from an albino and others from a coloured strain: subsequent histological examination of the pigmented retinas of the foetuses and newborn animals yielded morphological evidence of chimaerism (Tarkowski, 1963, 1964*b*). A similar method devised independently by Mintz was mentioned briefly in 1962 (Mintz, 1962*a, b*) and described more fully two years later (Mintz, 1964*a*). The techniques used were in some ways more convenient than those of Tarkowski (see Chapter 2), and embryos were aggregated not merely in pairs, but in groups of up to ten, forming giant blastocysts. Adult chimaeric mice were described first by Mintz (1965*a, b*) and later by Mystkowska & Tarkowski (1968) and others.

The term 'chimaera' was used by Tarkowski (1961). Mintz (1967*a*) coined an alternative term, 'allophenic', to refer to animals derived from embryo aggregation; however, Hadorn (1945, 1961) had previously used the terms 'allophene' and 'allophenic' in a different sense, referring to any characteristic of a cellular system which derives not from its own genetic constitution, but from the genetic constitution of some other cellular system (see discussion after McLaren, 1972*a*). Chimaeras made between embryos from different parents are often referred to as 'tetraparental' animals.

In earlier papers the embryos were described as 'fused' rather than 'aggregated', and the resulting animals were often referred to as 'fusion chimaeras'. This terminology has now been largely abandoned, in view of the potential confusion with virus-mediated cell fusion, and the expression 'aggregation chimaera' has been substituted. A $\leftrightarrow$ B denotes a chimaera formed by aggregating a strain A with a strain B embryo.

An alternative technique for making mammalian chimaeras has been devised by Gardner (1968). Cells from a dissociated blastocyst are injected singly or in groups into the blastocoele of an intact blastocyst, and become incorporated into the inner cell mass of the recipient embryo, giving rise to an animal that may be termed an 'injection chimaera'. Details of the technique are given in the next chapter. A $\rightarrow$ B denotes a chimaera formed by injecting strain A cells into a strain B blastocyst.

Methods for producing mammalian chimaeras have all been worked out on mouse embryos, for which much is known of the requirements for growth *in vitro* during the pre-implantation period, and have been slow to spread to other species. An attempt was made to produce aggregation chimaeras in sheep (Pighills, Hancock & Hall, 1968), but the only lamb thought from its transferrin type to be tetraparental was stillborn. More recently, injection chimaeras have been successfully produced in sheep (Tucker, Moor & Rowson, 1974) and rabbits (Gardner & Munro, 1974; Moustafa, 1974), and chimaeric rats have been made by the aggregation method (Mayer & Fritz, 1974; R. J. Mullen, personal communication).

No problems have been encountered in inducing aggregation between cleaving embryos of different strains, or even of different rodent species. Mouse–rat aggregation chimaeras develop successfully to the blastocyst stage (Mulnard, 1973; Stern, 1973; Zeilmaker, 1973), and cells from both species can be seen contributing to both inner cell mass and trophoblast. The dual origin of the trophoblast might be expected to interfere with the implantation process in either mouse or rat uteri, since blastocysts of neither species are able to implant normally in the uterus of the other (Tarkowski, 1962), and indeed no successful post-implantation development of mouse–rat aggregation chimaeras has been reported. Using the mouse–bank vole combination, Mystkowska (1975) attempted to produce aggregation chimaeras with trophoblast of entirely mouse origin by sandwiching one bank vole embryo between two mouse embryos. After transfer to a mouse foster-mother, some of the chimaeric embryos continued to develop for a few days, though morphological abnormalities were seen.

The most successful between-species experiments have been those of Gardner & Johnson (1973), using the injection technique to introduce rat inner cell masses into mouse blastocysts, before transfer to mouse foster-mothers. Evidence of the composite origin of embryos up to the egg cylinder stage (7 days of gestation) was obtained by treating alternate sections with fluorescent anti-rat and anti-mouse immune sera. Electrophoretic examination of later stages for enzyme markers suggested that the rat component decreased in amount towards the end of gestation (M. H. Johnson, personal communication). No proven chimaeras have as yet survived beyond birth.

Uses of chimaeras

This book will be mainly concerned with two types of experimental study for which chimaeras provide uniquely suitable material. The first belongs to the field of experimental embryology, making use of the two cell populations to trace the origin and fate of tissues and cell lineages in development, as earlier embryologists used cell populations marked with vital dyes. The second is an aspect of developmental genetics, and seeks to analyse how genetically different cells collaborate to form an adult animal. Although the distinction between the two approaches is often blurred, Chapters 3 and 4 will concentrate on the experimental embryology aspects, and Chapters 5 to 9 on developmental genetics.

The construction of 'fate maps', and the clonal analysis of development, have been pursued in an elegant manner in insects, especially *Drosophila*, using genetic mosaics rather than chimaeras (see for example Hotta & Benzer,

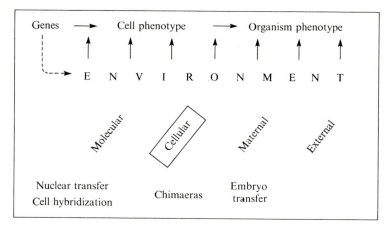

Fig. 4. Possible levels of interaction between genotype and environment. For explanation see text.

1972; Garcia-Bellido, Ripoll & Morata, 1973). The problems that this line of work faces are much greater in mammals than in insects, for a variety of reasons. In mammals cells move around and mingle during development, they interact with one another, and few adequate cell-localized markers are known. The problem of markers will be discussed more fully in Chapter 2. In *Drosophila*, on the other hand, very little cell mingling occurs, so that a clone usually remains as a coherent group of cells; most characters are cell-autonomous, in that there is rather little intercellular interaction; and the ubiquitous bristles form admirable cell markers, since each bristle is formed by a single cell, many genes are known that affect the structure and colour of bristles, and they are easy to classify.

On the other hand, the interactions between cells that make mammalian material less than ideal for the tracing of cell lineages are of particular interest when we come to consider how different cell populations collaborate. The old 'nature–nurture' problem (see Fig. 4) was concerned with how an organism of a particular genetic make-up interacts with its environment to achieve its final adult phenotype, and how much of this phenotype is determined by the genotype and how much by the environment. In mammals, the environment includes not only the external conditions to which the individual is subjected after birth, but also the maternal environment that it experiences before birth. The technique of embryo transfer can be used to investigate the importance of the maternal environment.

When we consider the development of individual cells, we are faced with a more sophisticated extension of the nature–nurture problem, since the environment in which a cell develops is made up of the population of other cells surrounding it. In a normal animal each cell is of course surrounded by others

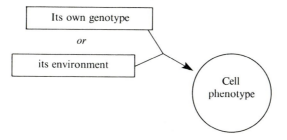

Fig. 5. The determination of cell phenotype.

of its own genotype, so questions of cell autonomy are hard to resolve; but in a chimaera we can ask whether the phenotype of an individual cell is determined by its own genotype or by its cellular environment (Fig. 5), we can study how such cells interact to form tissues and organs, and finally we can examine the phenotype of the whole animal. Thus for studies of the cellular environment, the technique of experimental chimaerism is the one of choice. For completeness, Fig. 4 includes also the cytoplasmic environment in which the nucleus and its genes operate; here the techniques of nuclear transfer and cell hybridization are relevant.

Experimental chimaerism has also been used as a genetical tool, to transmit the X-linked lethal mutation *jimpy* (*jp*) (Eicher & Hoppe, 1973). Aggregation chimaeras were made between embryos from *Ta jp*/ + + ♀ × + +/Y ♂ crosses and wild-type embryos, the viable X-linked *tabby* gene serving as a marker to identify overt chimaeras. One male chimaera was shown by breeding tests to transmit the *Ta jp* X chromosome, so enough normal cells must have been present to counteract the lethality of the *jp*/Y genotype and allow *jp*/Y cells to differentiate in the testis and form functional *jp* spermatozoa. This male produced abnormal offspring when mated to females heterozygous either for *jimpy*, or for another X-linked male lethal, *myelin synthesis deficiency* (*msd*). This established that the two 'lethal' genes were allelic, so that *msd* is now known as *jp^{msd}*. The use of chimaeras to maintain and study recessive lethals may enable numerous hitherto intractable problems to be solved.

2

Techniques

In this chapter I shall describe the techniques used for making experimental chimaeras in mammals, and also some of the techniques used to identify the different cell populations in the resulting composite animals. This account is not intended as a detailed practical guide to chimaera production: most of the relevant procedures are more fully described in Daniel (1971).

Making chimaeras

All experimental chimaeras so far described have been made either by aggregating cleaving embryos, or by injecting cells or groups of cells into blastocysts. Both techniques require the embryos to be manipulated *in vitro*, and subsequently transferred to the uteri of foster-mothers; embryo aggregation requires in addition that they be maintained *in vitro* for a period after treatment.

Embryo culture and transfer

Although some culture of mammalian (mainly rabbit) embryos had been carried out by Waddington and others in the 1930s, the cultivation of pre-implantation stages *in vitro* on anything like a routine basis had to wait until the late 1950s. By that time the introduction of antibiotics had greatly simplified all culture techniques, while the growing realization of the immense threat posed by the rate of world population increase was leading to greater emphasis on research in reproductive biology, including mammalian development. Following pioneer studies on mouse embryo culture by Hammond (1949) and Whitten (1956, 1957), a series of investigations by Brinster (e.g. 1965a, b) established the requirements for growing mouse embryos *in vitro* from the 2-cell or 8-cell to the blastocyst stage.

Rabbit embryos have also been extensively cultivated *in vitro*. Some success with culture of pre-implantation embryos has been achieved in other species, including rat, sheep and human, but knowledge of the biochemistry and nutritional requirements of embryos during the pre-implantation period is virtually limited to the mouse and rabbit.

A good account of mouse embryo culture, and the different culture systems

TABLE 1. *The composition of some culture media used for embryo aggregation*

Component (g/l)	Brinster, 1965a	Mullen & Whitten, 1971	Mulnard, 1971	Mystkowska & Tarkowski, 1970 (from Mulnard, 1967)	Mintz, 1964a, 1971a
NaCl	5.55	5.14	6.09	6.41	
$CaCl_2$	0.19	—	0.18	0.18	Equal parts
KCl	0.36	0.36	0.34	0.35	of foetal
KH_2PO_4	0.16	0.16	—	—	calf serum
$MgSO_4 \cdot 7H_2O$	0.29	0.29	0.114	0.09	and Earle's
$NaH_2PO_4 \cdot H_2O$	—	—	0.106	0.13	balanced
$NaHCO_3$	2.11	2.11	2.21	2.00	salt solution,
Na pyruvate	0.028	0.035	0.013	0.013	+0.1 %
Na lactate	2.42	2.42	1.38	1.33	lactic acid,
Ca lactate $\cdot 5H_2O$	—	0.53	—	—	0.002 %
Glucose	—	1.00	1.08	—	phenol red,
Crystallized serum albumin	1.00	3.00	—	5.00	adjusted with $NaHCO_3$
Calf serum (heat-inactivated)	—	—	20 %	—	(7.5 %) to pH 7.0
Penicillin	0.08	0.08	—	—	
Streptomycin	0.05	0.05	—	—	
Phenol red	—	0.01	0.02	0.02	
Gas phase	5 % CO_2 in air	5 % CO_2, 5 % O_2, 90 % N_2	5 % CO_2 in air	2 % CO_2 in air	5 % CO_2 in air
Culture system	Drops under oil	Culture tubes or drops under oil	Capillary tubes	Drops under oil	Special culture chambers

used in different laboratories, is given by Biggers, Whitten & Whittingham (1971). Probably the most widely used technique is that introduced by Brinster (1963) in which embryos are grown in small drops of culture medium under mineral oil in a plastic Petri dish. The medium consists of a bicarbonate-buffered salt solution of the Krebs–Ringer type, supplemented with energy sources (pyruvate, lactate, glucose), bovine plasma albumin, penicillin and streptomycin, and the cultures are maintained in an atmosphere of 5 % carbon dioxide in air. Table 1 gives details of the medium and culture system used in some of the laboratories which have been most active in producing chimaeras.

Embryo transfer from one female mammal to another was first successfully carried out by Walter Heape in 1890, using rabbits. More than sixty years elapsed before the technique began to be widely exploited, but it has now been used successfully in a wide range of laboratory and farm animals. The importance of synchrony between donor and recipient was established by Chang (1950) in rabbits and by McLaren & Michie (1956) in mice. The first demonstration of the normality of cultured embryos was provided when McLaren & Biggers (1958) grew mouse embryos from the 8-cell to the blastocyst stage *in vitro* before transferring them to the uteri of foster-mothers, where they developed to term and subsequently grew into healthy, fertile adult mice.

Most techniques of embryo transfer involve picking up the embryos in a small volume of fluid with a fine glass micropipette, and transferring them surgically into the ovarian end of the uterine horn of a pseudopregnant or genetically marked pregnant female at a stage of gestation similar to or 24 hours behind the stage of the transferred embryos. Trauma to or handling of the uterus has to be minimized, otherwise implantation is inhibited. A useful review of the different methods in common use, and the technical hazards, is given by Dickmann (1971). A non-surgical method of embryo transfer, giving a reasonably high success rate, has recently been described for the mouse (Marsk & Larsson, 1974), but it has not yet been used for chimaeric embryos.

Embryo aggregation

The initial experiments of Tarkowski (1961, 1963) involved removing the zona pellucida mechanically, by sucking the 8-cell embryo into a glass pipette of internal diameter just less than the diameter of the embryo. The zona-free embryos were placed in pairs in drops of culture medium under mineral oil and the fluid sucked out of the drop, using an even finer glass pipette, until the two embryos were pressed together by the walls of the drop. More culture medium was again introduced into the drop after 5–30 minutes, by which time the embryos had stuck together and aggregation was proceeding.

The procedure was a technical *tour de force*, to a degree that can only be fully appreciated by those who have themselves attempted to repeat it. Fortunately an easier and more convenient technique was described by Mintz (1964a, 1967b, 1971a), in which the zona pellucida was removed by treatment with the enzyme pronase (Mintz, 1962c) and the zona-free embryos merely placed in contact with one another at 37 °C. Later studies have all involved the use of pronase to remove the zona pellucida. Care must be taken not to expose the embryos to pronase for longer than is required for zona lysis. If the embryos are gently pushed together, aggregation will ensue even if the temperature is not maintained at 37 °C (Bowman & McLaren, 1970). Pushing

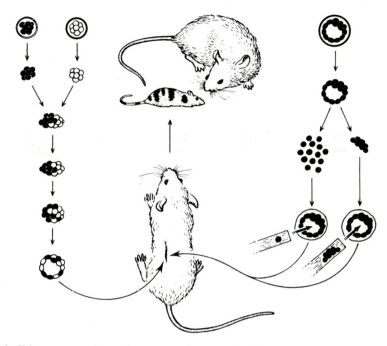

Fig. 6. Diagram to show the essential steps involved in making aggregation chimaeras (on the left) and injection chimaeras (on the right). For both, the procedures represented in the outer columns are carried out *in vitro*.

may be done with glass needles, forceps or hair loops. Mintz, Gearhart & Guymont (1973) describe the use of phytohaemagglutinin to facilitate aggregation at room temperature.

Once stuck, embryos continue to aggregate until a single large morula has been formed. This process has been filmed by time-lapse cinematography (Mulnard, 1971). The embryos are maintained in culture for 24–48 hours, to allow aggregation to be completed and to check that no blastomeres are extruded. They are then transferred to the uterus of a pseudopregnant foster-mother, $2\frac{1}{2}$ days *post coitum* (*p.c.*). Mintz uses special culture dishes, with a medium containing serum; Whitten uses tubes, in an atmosphere of 5% carbon dioxide, 5% oxygen and 90% nitrogen; and Mulnard places his embryos in drops of medium in capillary tubes immersed in oil, with the gas phase provided by bubbles in the capillary tube. Other published studies are based on the culture system of Brinster, with drops of modified Krebs–Ringer bicarbonate under mineral oil.

The essential steps of the procedure are presented in Fig. 6, and details of some of the culture media in Table 1. In our laboratory (Bowman & McLaren, 1970), blastocysts formed after zona removal and aggregation showed as high a percentage of successful development after transfer as did those that developed in culture without any manipulation, and foetal growth was also unaffected; on the other hand, cultured blastocysts, aggregated or not, proved less viable and gave rise to somewhat smaller foetuses than did blastocysts taken from the uterus. Mullen & Carter (1973) found that pronase treatment reduced the percentage of embryos developing successfully, but aggregation had no further deleterious effect. They recommend transfer of embryos to both uterine horns rather than to one only.

Aggregation is possible at any stage of cleavage. Although the 8-cell stage (about 60 hours *p.c.*) is usually the most convenient, embryos have been used at the 2-cell (Mintz, 1962*b*; Mulnard, 1971) and 4-cell (Hillman, Sherman & Graham, 1972) stage, or as late morulae (Mintz, 1964*a*). Most workers find that embryos will no longer aggregate once blastocyst formation has begun; Stern & Wilson (1972) report aggregation of 'late morulae/early blastocysts' at 72 hours *p.c.*, when the mean cell number would not much exceed 16.

Aggregation chimaeras can be made between more than two entire embryos (e.g. Mintz, 1965*a*, up to 16 embryos; Hillman *et al.*, 1972, 15 embryos), as well as between various numbers of isolated blastomeres or parts of embryos (Hillman *et al.*, 1972).

Injection chimaeras

A more difficult technique, but one that offers the opportunity of studying a number of problems inaccessible to aggregation-chimaera methods, was described by Gardner (1968). The essential steps are shown in Fig. 6.

After removal of the zona pellucida with pronase, donor mouse blastocysts are disaggregated in calcium- and magnesium-free saline or versene solution and one or more cells are picked up in a fine glass pipette. The recipient blastocyst is held on a suction pipette and, using micromanipulators, a triangular hole is made in the trophoblast wall by introducing, and then separating, three very fine glass needles. The donor cells are introduced into the blastocyst cavity by inserting the injection pipette through the hole. The blastocysts are transferred to the uterus of a foster-mother. Since the mice that are born are often visibly chimaeric, the donor cell or cells must become incorporated in the inner cell mass of the recipient blastocyst.

An essentially similar technique, but using two rather than three micro-needles, has been used by Moustafa & Brinster (1972*a*) to investigate the fate of heterochronic cells transferred to mouse blastocysts. Comparison with control blastocysts suggested that the procedure did not affect the blastocyst's chances of survival.

When a whole extra inner cell mass is surgically transferred from one mouse blastocyst to another (Gardner, 1971), aggregation with the resident inner cell mass usually occurs, giving rise to a chimaeric embryo.

The injection technique can be used in situations for which aggregation methods are inappropriate, e.g. for species such as the rabbit (Moustafa, 1974) where removal of the zona pellucida precludes subsequent development, or for making between-species chimaeras (Gardner & Johnson, 1973), where a double contribution to the outer trophoblast layer might interfere with normal implantation in the host uterus.

Cell markers

The ideal cell marker to distinguish one chimaera component from the other would be cell-localized, cell-autonomous, stable, distributed universally among both the internal and the external tissues of the body, and easy to detect, both grossly and in histological sections, without elaborate processing. It would have several genetically determined variants, characterizing different strains. No such marker exists.

Early development

For the analysis of blastomere distribution during early development, use has been made of genetic differences in granularity of the cytoplasm, and of the abnormal appearance of homozygous lethal t^{12}/t^{12} embryos (Mintz, 1964a). Tritiated-thymidine labelling of one component has also been used as a marker (Mintz, 1964a; Garner & McLaren, 1974), but no attempt has yet been made to apply techniques of vital staining, as used for example by Daniel & Olsen (1966) on rabbit embryos.

Antigenic markers

Antigenic variation comes closest to the ideal, and has been exploited in a number of different ways. Histocompatibility antigens potentially satisfy all the requirements listed above, since they are cell-localized, cell-autonomous, present in all tissues, and possessed of a virtually infinite amount of genetic variation. The only problems arise in their detection. Red cell antigens are easy to detect by haemagglutination tests, so can be used to register and quantitate erythrocyte chimaerism in strain combinations differing at appropriate histocompatibility loci (Mintz & Palm, 1965c, 1969; Wegmann & Gilman, 1970). Lymphocytes can be typed, using theta antigens (Bona, Tuffrey & Barnes, 1974). Skin grafting onto recipients of the component strains has been used in an attempt to study chimaerism in hair follicles and melanocytes (Mintz & Silvers, 1970), but the interpretation of partial graft breakdown is difficult. Immunofluorescent methods for antigen detection of tissue sections offer the most promising approach to localization at the cellular level, but unfortunately the techniques are complicated and often prove unreliable. Weak histocompatibility antigens, such as the male antigen, H-Y, cannot at present be localized by immunofluorescent techniques; strong histocompatibility differences, at the *H-2* locus, have been successfully used in some laboratories (Barnes, Holliday & Tuffrey, 1974); perhaps the most valuable application of such techniques so far is in the analysis of inter-species chimaeras (Gardner & Johnson, 1973).

Biochemical markers

Other biochemical variations have also proved valuable. Erythrocyte chimaerism can be studied using haemoglobin variants (Wegmann & Gilman, 1970) as well as antigenic markers. Serum proteins (e.g. transferrins) can be typed (Pighills, Hancock & Hall, 1968). Genetically determined enzyme variation, detectable by starch or polyacrylamide gel electrophoresis, has been widely used, particularly with reference to isocitrate dehydrogenase (IDH) and glucose phosphate isomerase (GPI). These enzymes are distributed in most tissues of the body, so with their aid chimaerism can be detected in liver, kidneys, gonads, lungs, heart, spleen, brain, blood (not IDH), skeletal muscle and placenta (both foetal and maternal components). The test for GPI is particularly sensitive, and can detect a minor component at the 1 % level (Chapman, Ansell & McLaren, 1972) using very small amounts of tissue, such as the ectoplacental cones of individual mouse embryos 8 days *p.c.* (Chapman *et al.*, 1972), or single somites from 10-day embryos (Gearhart & Mintz, 1972*a*). Both IDH and GPI are dimers, so that the presence or absence of a 'hybrid band' on the gel can be used as a test of cell fusion in a chimaeric tissue (Mintz & Baker, 1967; Chapman *et al.*, 1972; Gearhart & Mintz, 1972*b*; Ansell, Barlow & McLaren, 1974).

The chief drawback of the electrophoretic approach is that it requires homogenization of the tissues, so that no information can be derived as to the spatial distribution of the chimaera components at the cellular level. This problem could be overcome by the use of histochemical techniques, but so far the only enzyme variation detectable histochemically in the mouse is an activity difference affecting β-glucuronidase (Condamine, Custer & Mintz, 1971; West, 1976*c*), and this is of limited usefulness since it appears to be manifest only in the liver (Fig. 7).

Liver chimaerism can also be diagnosed in living animals by electrophoresis of urine, using a combination of strains that differ at the *Mup-1* locus (*major urinary protein*) (Baker & Mintz, cited by Condamine *et al.*, 1971).

Chromosomal markers

Chromosome variations between strains (e.g. translocations) provide a useful and technically straightforward approach to determining the relative pro-

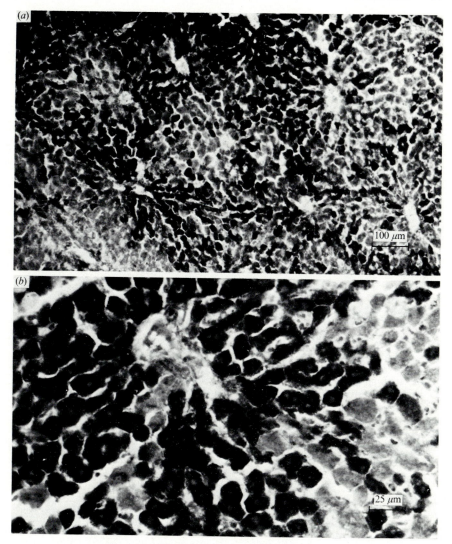

Fig. 7. Chimaerism for activity levels of β-glucuronidase in the liver of a C3H↔C57BL mouse. (*a*) is a low power and (*b*) a higher power view of a section of liver. The pale cells belong to the C3H component, and stain light green, indicating low β-glucuronidase activity; the darker cells are genetically C57BL, staining red on account of their high β-glucuronidase activity. (Reproduced by kind permission of John West.)

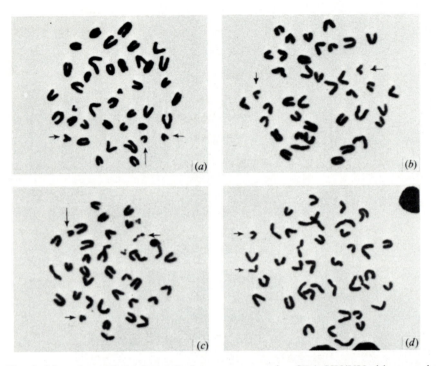

Fig. 8. Metaphase plates from the bone marrow of a CBA XX/XY chimaera, in which the XY component carried two T6 marker chromosomes. (*a*) and (*c*) show CBA-T6T6 male plates, each with a Y chromosome (long arrow), the two shortest autosomes (short arrows), and two even smaller T6 marker chromosomes. (*b*) and (*d*) show CBA-p plates of female genetic sex, with no chromosomes smaller than the two shortest autosomes (arrowed). From Mystkowska & Tarkowski (1968).

portions of the two components in a variety of different tissues. The variant used most widely for this purpose is the T6 chromosome translocation (Fig. 8): strains homozygous for the translocation have two small 'marker' chromosomes, while heterozygotes have a single such chromosome (see e.g. Mystkowska & Tarkowski, 1968; Ford, Evans & Gardner, 1975*b*). Variations between strains in the amount of pericentric heterochromatin on certain chromosome pairs, visualized by Giemsa staining (Fig. 9), have been similarly used (McLaren, 1975*a*), and in the special case of XX/XY chimaeras, chromosomal sexing with the aid of either Giemsa staining or autoradiographic detection of the Y chromosome (Fig. 10) provides yet another means of discriminating one component from another (McLaren, 1972*c*).

Chromosomal techniques have two major disadvantages: they cannot be used on histological sections, so cannot provide any information on cell localization within tissues, and they can be applied only to cells undergoing

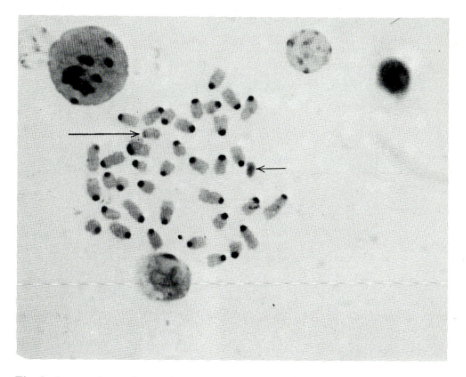

Fig. 9. A metaphase plate stained by a Giemsa C-banding technique, from a male F₁ between strains polymorphic for the amount of pericentric heterochromatin on chromosome 14. The Y chromosome, darkly staining and with no pericentric heterochromatin, is indicated with a short arrow; the 'marker' chromosome 14, showing a very much reduced amount of pericentric heterochromatin, is indicated with a long arrow.

division. If the cells of the chimaera components have different mitotic rates in a particular tissue, chromosome techniques will yield a biassed estimate of the relative proportions of the two cell types in that tissue. In addition, although a few adult tissues (bone marrow, spleen, cornea) contain enough dividing cells to yield useful information from direct chromosome preparations, most tissues require to be grown for several days in culture before chromosome preparations are made, giving the possibility of still more biassed estimates if the chimaera components respond differently to in-vitro culture or, in the case of lymphocytes, to the stimulating effect of phytohaemagglutinin.

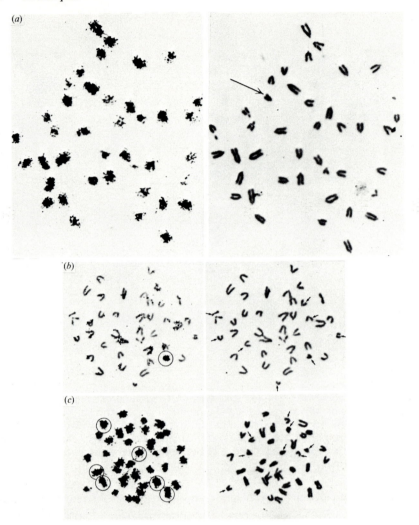

Fig. 10. Autoradiographic detection of the Y chromosome. In each pair of photographs, an autoradiograph is shown on the left, and on the right the same photograph after removal of the silver grains. (*a*) shows a very well-spread plate, in which the Y chromosome (arrowed) could be identified on morphology alone: note that it is heavily labelled. (*b*) and (*c*) are less good plates; the most heavily labelled chromosomes are ringed in the pictures on the left, and on the right the eight smallest chromosomes are arrowed. Only one chromosome falls into both categories: this is the Y chromosome.

Pigment

A genetic variant, stable and easy to detect without any special processing, is presence or absence of melanin pigment. Melanin is cell-localized in the pigmented retina of the eye (Tarkowski, 1963, 1964*b*; Deol & Whitten, 1972*a*) and in certain areas of the inner ear (Deol & Whitten, 1972*b*); in these locations it provides an ideal cell marker and gives an accurate picture of the distribution of the genetically distinct cells. Unfortunately in the choroid layer of the eye and in hairs the pigment is extracellular, so that its distribution only approximately reflects that of the progenitor melanocytes. As will be seen in Chapter 5, a hair follicle may be colonized by several melanoblasts, so that a single hair in a chimaera may be of mixed phenotype. Further, pigment production in hair follicles is not always cell-autonomous, since for certain genetic variants (e.g. agouti/non-agouti) the phenotype of the melanocyte depends on the genotype of the hair follicle that it has colonized (see Chapter 5).

Other morphological markers

Histological and ultrastructural studies by Drews & Alonso-Lozano (1974) suggest that response to androgens behaves as an autonomous character at the cellular level in the epididymis of sex-reversed mice (i.e. phenotypic males XX in chromosome constitution) mosaic for the X-linked *testicular feminization* gene (see Chapter 9 for further details of X-inactivation mosaicism). Few other morphological variants have yet been studied at a cellular level, so as to provide information on the cellular distribution of the two genetic components. A possible exception is *rd*, a gene producing degeneration of the neural retina (Mintz & Sanyal, 1970; Wegmann, LaVail & Sidman, 1971). Partial localization is seen in some systems, such as the distribution of hairs and epidermal plaques on the tails of tabby chimaeras (McLaren, Gauld & Bowman, 1973), and the shape of vertebrae (Moore & Mintz, 1972). Other morphological markers can merely act as indicators of chimaerism by showing intermediate expression, as for example the short or malformed tails sometimes seen in chimaeras made from homozygous vestigial tail and normal embryos (see Chapter 6).

3

Early development

This chapter is concerned with experimental embryology, with the pre-implantation and early post-implantation stages of development of the mammalian (in effect mouse) embryo, and with the light that studies of chimaeric embryos have been able to throw on the underlying causal basis of the developmental events taking place during this period. Table 2 gives the timing of some of the landmarks of mouse development that will be discussed.

The very fact that embryo aggregation is possible in the mouse, and that it results in a single adult individual of normal size and shape rather than any form of double monster, bears witness to the astonishing regulative powers of mammalian embryos. This aspect was stressed in the early papers on aggregation chimaeras (Tarkowski, 1961, 1965; Mintz, 1962*b*, 1964*a*, *b*, 1965*a*); it has been fully borne out by later work. Tarkowski (1961) thought that the developmental lability of the blastomeres of the mouse embryo was lost during the mitotic cycle following the 8-cell stage, but subsequent experiments showed that successful aggregation could be obtained between 16-cell and even 32-cell morulae (Mintz, 1964*a*). Stern & Wilson (1972) report some aggregation between early blastocysts, but these were probably still at the 16-cell stage, as they were only 12 hours older than the 8-cell embryos used in the same study. Reaggregation of later stages following complete dissociation in versene has been reported (Stern, 1972); possibly the versene treatment alters the properties of the cell surface. Heterochronic aggregations and reaggregations, between embryos or blastomeres of different ages, are also possible (Mintz, 1962*b*; Mulnard, 1971; Stern & Wilson, 1972; Stern, 1972), but the proportion developing successfully to the blastocyst stage is lower than when the two component embryos are the same age as one another. The participation of heterochronic partners in post-implantational development has not been examined.

The developmental potential of individual blastomeres at the 4-cell stage has been investigated by Kelly (1975) in the mouse. After dissociation, each of the four blastomeres was surrounded by 'carrier' cells from another embryo, distinguishable both by coat colour genes and by an enzyme marker. When all four composite embryos successfully continued their development after transfer to a uterine foster-mother, each blastomere was found to have contributed to the foetus, to the yolk sac, and to the placenta; in several cases

TABLE 2. *The timing of development in the mouse**

Event	Days after fertilization
2-cell stage	$1\frac{1}{2}$
8-cell stage	$2\frac{1}{2}$
Embryos enter uterus	3
Differentiation of trophectoderm and inner cell mass	$3-3\frac{1}{2}$
Blastocyst stage	$3\frac{1}{2}$
Implantation begins	4
Differentiation of primary endoderm and ectoderm	4–5
Appearance of proamniotic cavity	$5\frac{1}{2}-6$
Appearance of primitive streak	$6\frac{1}{2}-7$
Somite formation begins	$8-8\frac{1}{2}$
Formation of chorioallantoic placenta	9–10
X chromosome inactivation not yet detectable, by genetic criteria	$4\frac{1}{2}$
X chromosome inactivation detectable, by late DNA replication	6–7
Primordial germ cells first seen, near base of allantois	$7\frac{1}{2}-8$
Primordial germ cells enter genital ridge	$10\frac{1}{2}-11$
Female germ cells commence meiosis	13–14
Melanoblasts leave neural crest	$8\frac{1}{2}-9\frac{1}{2}$
Haematopoiesis begins in yolk sac	$8\frac{1}{2}-9$
Haematopoiesis begins in foetal liver	$9\frac{1}{2}-10$
Haematopoiesis begins in foetal bone marrow	16–17
Birth	19–20
	Days after birth
Hairs erupt	6–7
Vaginal opening	28–35
Sexual maturity	35–42

* The times given are based on the randomly bred Q strain. Some inbred strains develop more slowly, with a total gestation period of up to 1 day more.

in which three of the four embryos were reared to adulthood, descendants of each single blastomere were found distributed among most organs of the body. A similar result was obtained when pairs of 8-cell blastomeres were followed into post-implantation development, suggesting that totipotency continues at least until the 8-cell stage.

Differentiation of inner cell mass and trophectoderm

It seems that formation of the fluid-filled blastocyst involves the differentiation of a new tissue, the outer trophectoderm layer, whose properties (including the appearance of intercellular 'tight junctions') both allow the accumulation

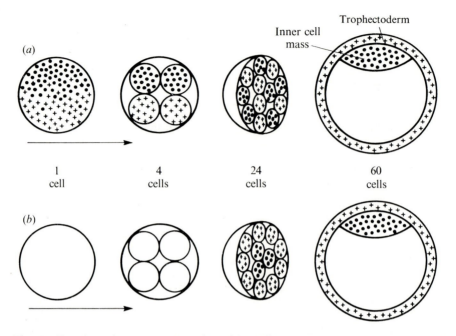

Fig. 11. Two hypotheses to explain the initial differentiation of trophectoderm and inner cell mass in the mouse blastocyst. (*a*) is a preformationist theory, in which the critical differences already exist in the cytoplasm of the fertilized egg; (*b*) is an epigenetic theory, in which the differences arise during cleavage, from the location of the blastomere within the embryo.

of fluid and, at the same time, prevent cell-to-cell aggregation. Gardner & Johnson (1972) found that trophectoderm fragments isolated from mouse blastocysts would not stick together; in contrast, isolated inner cell masses readily aggregate, either with one another, or with 8-cell embryos (R. L. Gardner, personal communication). The basis for this first process of differentiation, into inner cell mass and trophectoderm, has been investigated using aggregation methods.

Dalcq and his co-workers (see Dalcq, 1957) proposed that differentiation of the early mammalian embryo arises as a result of some pre-existing heterogeneity in the cytoplasm of the fertilized egg, which becomes parcelled out into separate cells during the course of cleavage (Fig. 11*a*). Mintz (1965*a*), on the other hand, concluded from her studies on aggregation chimaeras that any rigid 'pre-patterning' must be considered unlikely, at least in the mouse. Aggregates of up to sixteen mid-cleavage embryos successfully formed blastocysts at the normal time, suggesting perhaps that no 'sorting out' of predetermined inner cell mass and trophectoderm cells was required. Pairs of embryos were aggregated which could be distinguished from one another, either by autoradiography after tritiated-thymidine labelling of one partner, or through the characteristic granularity of the cytoplasm of one strain, or by taking one embryo of the homozygous lethal t^{12}/t^{12} genotype (Mintz, 1964a). The results were not conclusive, but no evidence of any 'ordered deployment' of the two types of cells was seen.

Later work, in which isotopically labelled and unlabelled 8-cell embryos were aggregated and the resulting blastocysts serially sectioned and subjected to autoradiography, confirmed that little or no cell mingling takes place up to the blastocyst stage (Garner & McLaren, 1974).

The proposal that location within the embryo, either internal or external, at a particular stage of development, might be the factor determining whether a particular blastomere differentiated as inner cell mass or trophectoderm, was put forward by Tarkowski & Wroblewska (1967) as a result of their experiments on the isolation of blastomeres at the 4- and 8-cell stage. They suggested that any cell on the outside after three to four cleavage divisions (normally, the 8–16-cell stage) differentiated as trophectoderm, while those on the inside formed the inner cell mass (Fig. 11b). If cell number were experimentally reduced at the critical period, all cells would be on the outside, so that a 'trophoblastic vesicle' would be formed, lacking an inner cell mass completely.

This hypothesis was strongly supported by the elegant recombination experiments of Hillman *et al.* (1972). To test the effect of 'outside' location, 4- and 8-cell mouse embryos were labelled with tritiated thymidine, then disaggregated and each blastomere stuck onto the outside of a different unlabelled zona-free 4- to 16-cell embryo. Autoradiography at the blastocyst stage revealed that 90 % of the labelled cells were in the outer trophectoderm layer; in no embryo were the labelled cells found only in the inner cell mass, and in the case of three donor embryos all the labelled derivatives were in the outer layer, although had they been left *in situ* some of their descendants would of course have formed the inner cell mass. When a similar experiment was performed with donor and host differing by an electrophoretically detectable variant of the enzyme GPI, descendants of the donor blastomeres were always present in the trophoblast and/or yolk sac on the 10th day of gestation, and only in four out of twelve cases did they form part of the embryo proper. Conversely, when labelled 8-cell stages were entirely surrounded with fourteen unlabelled embryos (Fig. 12), the labelled cells were found only in the inner cell mass in four out of seven attempts, and in both inner cell mass and trophectoderm in the remaining three. After transfer, such giant chimaeric blastocysts proved capable of developing at least until the 13th day of gestation.

The conclusion that the location of a blastomere during cleavage could determine the direction of its differentiation was underlined by the study of Wilson, Bolton & Cuttler (1972). Droplets of silicone fluid injected as

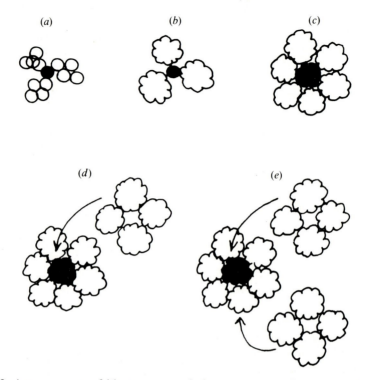

Fig. 12. Arrangements of blastomeres and cleavage stage embryos in such a way that labelled cells (black) were located in the interior of the cell mass, partially or completely surrounded by unlabelled cells (white). The most successful combination is shown in (*c*), (*d*) and (*e*): a labelled 8-cell embryo is surrounded by six unlabelled embryos (*c*), a cap of four similar embryos is placed on top (*d*), and a further cap of four embryos on the bottom (*e*). From Hillman, Sherman & Graham (1972).

markers into the outer areas of blastomeres of cleaving (2-cell to morula) mouse embryos were found at the blastocyst stage exclusively in trophectoderm cells, never in the inner cell mass. However, when such embryos were aggregated, so that some of the droplet-containing blastomeres were apposed, the droplets were found in the inner cell mass of the composite blastocyst, as well as in the trophectoderm.

Subsequent fate of trophectoderm and inner cell mass

In normal mouse embryos the trophectoderm adjacent to the inner cell mass (polar trophectoderm) is believed to give rise, by continued proliferation, to the ectoplacental cone; the rest of the trophectoderm (mural trophectoderm) undergoes no further cell division after implantation, but instead transforms into giant cells. When inner cell mass and polar trophectoderm are cut off, the remaining trophectoderm rounds up into a trophoblastic vesicle, which is capable of inducing a decidual reaction but shows no further cell division (Gardner & Johnson, 1972). If, however, a substitute inner cell mass is injected into such a trophoblastic vesicle, the reconstituted blastocyst develops normally, with proliferation of trophoblast to form a normal ectoplacental cone (Gardner, 1971).

The proliferating ectoplacental cone tissue could, of course, be derived directly from the inner cell mass, rather than from trophoblast under the influence of the inner cell mass. This possibility was tested by recolonizing trophoblastic vesicles of one GPI type with inner cell masses of an electrophoretically distinct GPI type (Gardner, Papaioannou & Barton, 1973). When examined at the early somite stage the ectoplacental cone tissue turned out to be almost exclusively of the trophoblast donor type. The authors conclude that the inner cell mass makes little or no cellular contribution to the ectoplacental cone, but controls trophoblast proliferation, perhaps by inhibiting giant cell transformation in those trophoblast cells adjacent to it. On the other hand the embryo and extra-embryonic membranes, believed to be derived entirely from the inner cell mass, turned out also to include a substantial proportion of the trophoblast donor enzyme. Subsequent studies of mouse–rat composites, with a rat inner cell mass injected into a mouse blastocyst, suggest that the trophectoderm-derived component is in fact the extra-embryonic ectoderm (Gardner, 1975b).

The development of composite mouse–rat embryos has been followed up to the 30-somite stage (Gardner & Johnson, 1973, 1975; Johnson & Gardner, 1975; Gardner, 1975a). From fifty-six chimaeric blastocysts transferred, twenty-eight living embryos were recovered. Chromosomal analysis showed chimaerism in the first four embryos tested; a further seventeen were therefore sectioned and treated with fluorescent antisera, with anti-mouse and anti-rat applied to adjacent serial sections. Five showed convincing evidence that rat cells as well as mouse were present in the endoderm, and three of these had both rat and mouse cells in the mesoderm and ectoderm also. Positive staining for rat antigens in one section was associated with deficiency of staining for mouse antigens in corresponding areas of the adjacent section, and vice versa (Fig. 13). Although the rat cells were transferred as a clump, they were detected subsequently as several discrete patches, indicating that cell migration had occurred. Some of the patches were large, so some coherent clonal growth must have occurred, and the rat patches in one germ layer showed no obvious spatial relationship to those in the adjacent cell layer. Rat cells from the injected inner cell mass contributed not only to the embryo, but also to the amnion, distal endoderm and yolk sac.

Chimaeric embryos pose a problem in growth regulation, since 'double' blastocysts formed by aggregating two 8-cell stages contain twice the normal

3

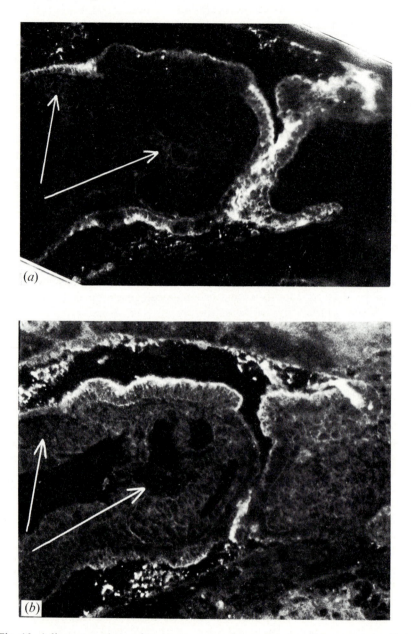

Fig. 13. Adjacent sections of a rat→mouse embryo, showing chimaerism in all three germ layers. (*a*) is stained with antiserum to rat species antigens: two mesodermal patches (arrowed) are staining positively. (*b*) is stained with antiserum to mouse species antigens: the two mesodermal patches positive in (*a*) now show a deficiency of staining. From Gardner & Johnson (1973).

number of cells (Buehr & McLaren, 1974,) yet by late gestation foetal and placental weight is indistinguishable from that of control 'single' embryos (Bowman & McLaren, 1970). Tarkowski (1963) reported that 'double' embryos 'fall within the normal range of variation in size' from $6\frac{1}{2}$ days *p.c.* onwards; Mintz (1971*b*) states: 'normal embryo size is restored during implantation', perhaps by 'origin of the embryo itself from a small fixed number of cells...in the inner cell mass, regardless of the total cell number in the blastocyst'.

In the normal mouse embryo differentiation of endoderm is followed by downgrowth of the egg cylinder into the blastocyst cavity, while within the egg cylinder segregation of embryonic and extra-embryonic ectoderm occurs. At about $5\frac{1}{2}$ days *p.c.* the proamniotic cavity is first seen. Volume estimates from serial sections (Buehr & McLaren, 1974) indicate that throughout this period the egg cylinders of 'double' embryos are at least twice as large as those of 'single' embryos. At later stages, however, no size difference between 'double' and 'single' egg cylinders could be detected, suggesting that, in accordance with the observations of Tarkowski (1963), growth regulation takes place $5\frac{1}{2}$–6 days *p.c.* Since regulation appears to involve the entire egg cylinder, it is unlikely to be due to the embryo originating from a small fixed number of cells.

Cell fusion during development

Some cell types in the body contain substantially more than the 2c–4c amount of DNA characteristic of normal diploid cells. The additional DNA may all be in one large nucleus, or may be distributed among several nuclei. Two developmental paths could lead to such a situation: fusion between cells, or continued replication of the DNA within a cell ('endoreduplication'), with or without nuclear division, but without cell division. The occurrence of fusion can be detected if the cells concerned differ with respect to electrophoretic variants of a dimeric enzyme, as synthesis of both forms of the enzyme in a single cell (for example, in heterozygotes) yields an intermediate 'hybrid' band on gel electrophoresis. One way of achieving a genetically mixed population of cells for this purpose is by making chimaeras.

This principle was first applied by Mintz & Baker (1967) to investigate the developmental history of the multinucleate myotubes found in skeletal muscle, using the dimeric enzyme, NADP-dependent isocitrate dehydrogenase (IDH). Aggregations were made between embryos of strains C3Hf (*Id-1ᵃ/Id-1ᵃ*) and DBA/2 (*Id-1ᵇ/Id-1ᵇ*), and C57BL/6 (*Id-1ᵃ/Id-1ᵃ*) and CBA (*Id-1ᵇ/Id-1ᵇ*). Various tissues of the adult chimaeras were homogenized and analysed by starch gel electrophoresis: two bands, corresponding to the two enzyme variants *Id-1ᵃ* and *Id-1ᵇ*, characterized homogenates of chimaeric liver, kidney, lung, spleen and cardiac muscle, but skeletal muscle yielded in

3-2

addition the intermediate band that indicated the presence of hybrid enzyme. The authors concluded that the multinucleate myotubes of skeletal muscle must arise *in vivo* by fusion of separate myoblasts, as they were already known to do *in vitro*.

The absence of hybrid enzyme in other chimaeric tissues implies that, in heterozygotes, both alleles must be active in the same skeletal muscle cell, since assembly of hybrid molecules evidently cannot take place by diffusion of subunits across cell boundaries. Its presence in multinucleate skeletal muscle myotubes implies that each subunit is separately coded for by the genetic material, with assembly taking place in the cytoplasm.

To investigate the genetic control of another enzyme, NADP-dependent malate dehydrogenase (MDH), Baker & Mintz (1969) again made chimaeras between strains differing for electrophoretically distinguishable variants of the enzyme, and examined homogenates of skeletal muscle and other tissues (liver, kidney, lung, salivary gland, mammary gland, cardiac muscle). Heterozygotes for the two MDH alleles show a five-banded pattern on electrophoresis, suggesting a tetrameric structure for the molecule, with two subunits A and B, giving three 'hybrid' bands, ABBB, AABB and AAAB. Skeletal muscle from the chimaeras showed a similar pattern to that seen in heterozygotes, but other tissues again showed only the two pure bands. This result not only confirmed the tetrameric structure, but indicated that the MDH molecule is built up, not from two dimers, but from monomeric subunits assembled independently in the cytoplasm.

Cell hybridization, followed by the restoration of diploidy, has been shown to take place among somatic cells *in vitro*, but enzymic and chromosomal studies (Mintz, 1970*a*) have so far yielded no evidence for any such process in normal mammalian development.

The largest cells in the body are probably trophoblast giant cells, the nuclei of which contain in the mouse up to a thousand times the haploid DNA content. Using a very sensitive electrophoretic assay system for glucose phosphate isomerase (GPI), it proved possible to determine the GPI-1 type of individual ectoplacental cones (trophoblast) as early as the 8th day of pregnancy, and to detect a minor component constituting 1 % or more of the total GPI-1 activity of the sample (Chapman *et al.*, 1972). When 8-cell stages from strains differing at the *GPI*-1 locus were aggregated, two enzyme bands, indicating chimaerism, were found in embryos, embryonic membranes and trophoblast at the 11th day of pregnancy, but no trace of a hybrid band was detected. Trophoblast outgrown from chimaeric blastocysts *in vitro* also failed to show any hybrid enzyme. It was concluded that trophoblast cells do not fuse with one another either *in vivo* or *in vitro*, and that the increase in DNA probably occurs by endoreduplication. A similar result was obtained by Gearhart & Mintz (1972*b*) on trophoblast taken from chimaeric conceptuses on the 10th and 19th days of pregnancy.

The decidualized uterus is also characterized by binucleate cells and cells with up to sixty-four times the haploid amount of DNA. Electrophoretic analysis of GPI-1 types was therefore applied to chimaeric decidua on the 8th day of pregnancy (Ansell *et al.*, 1974). Again no hybrid enzyme could be detected electrophoretically, indicating that the increase in DNA content was due to DNA replication without cell division, rather than to cell fusion.

Injection chimaeras

The injection of two to five cells from dissociated blastocysts of a pigmented genotype (CBA-T6T6 × PDE) into the cavity of blastocysts from an albino strain (PDE) resulted in overt chimaerism, both at mid-gestation and after birth (Gardner, 1968). The proportion of chimaeras was low (less than 20 %), but in those animals that were chimaeric the injected cells had made a substantial contribution to the phenotype. At mid-gestation two foetuses showed strong pigmentation in the eye, and 15–20 % of donor-type metaphases (T6T6); of the two chimaeras that survived to weaning, one showed extensive pigmentation of the coat and underlying skin, while the other (from a reciprocal transfer of albino cells into a pigmented blastocyst) showed only limited chimaerism.

In a recent study (Ford *et al.* 1975*b*), a further two adult chimaeras of the same type (CBA-T6T6 × PDE) → PDE, were analysed. Identification of the T6 marker chromosomes showed that the contribution of the donor cells was remarkably high in a number of different tissues, including bone marrow, thymus, Peyer's patches, spleen, lymph nodes, gut, cornea (left and right), ovaries, testes, blood, skin and kidney. The proportions of donor cells ranged from 9 % (bone marrow) to 91 % (cornea) in one animal, and from 6 % (testis) to 68 % (blood) in the other. Since the reciprocal strain combination has not been studied, it is difficult to know to what extent the high incidence of donor cells reflects a selective advantage on the part of the donor strain, leading for example to a higher rate of cell division, and to what extent it implies that the total number of cells in the blastocyst contributing descendants to the foetus is normally very low. The distribution of the donor cells will be discussed further in Chapter 10.

The degree to which cells of a different developmental age can contribute to the embryo has been examined by Moustafa & Brinster (1972*a, b*). When cells from embryos 0, 4 and 8 days older were labelled with tritiated thymidine and injected into blastocysts, autoradiography revealed tht the donor cells were still detectable in the blastocysts after 40 hours' in-vitro culture, and indeed had increased somewhat in number. However, it seems that heterochronic cells are less likely to get incorporated in the developing embryo than are those of the same age. Use of a pigment marker showed that donor cells 0, 2, 4 and 8 days older than the recipient blastocysts gave rise

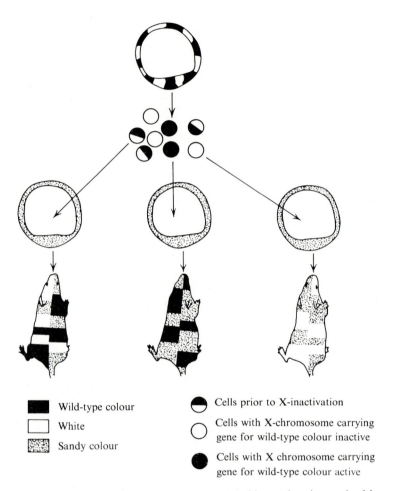

Fig. 14. A diagram to show the coat patterns expected in sandy mice made chimaeric by transfer of a single female donor cell from a female embryo heterozygous for an X–autosome translocation. The gene for wild-type colour (agouti) is carried on the normal X chromosome, and is therefore expressed only in cells in which this chromosome is active. If X-inactivation has occurred prior to transfer, each chimaera will show only two colours: if it occurs later, three colours may be seen. After Gardner, R. L., *Birth Defects and Fetal Development: Endocrine and Metabolic Factors*, ed. K. S. Moghissi, 1974. Courtesy of Charles C. Thomas, Publisher, Springfield, Illinois.

to chimaerism in 19 %, 15 %, 3 % and 0 % respectively of foetuses examined at 15–17 days' gestation; for postnatal development, the incidence of chimaerism (assessed by pigmentation of eyes, coat and progeny) was 15 %, 13 % and 0 % with donor cells 0, 2 and 4 days older than the recipients, respectively. Thus it seems that cells are most unlikely to participate in the development of an embryo if they deviate from it in age by more than 2 days.

Ingenious use was made of the possibility of injecting single cells into blastocysts, to investigate the time of X chromosome inactivation (see Chapter 9) in mice (Gardner & Lyon, 1971). The blastocysts which provided donor cells were from albino females mated to males carrying Cattanach's translocation: female progeny from such a mating would have coats mosaic for albino and agouti, corresponding to the inactivation of the translocated and normal X chromosomes respectively. The recipient blastocysts were from matings which would yield sandy-coloured progeny. Thus if any chimaeras were produced with both albino and agouti, as well as sandy, patches in the coat, it would indicate that the single injected cell was genetically female, and had not yet undergone X chromosome inactivation, so that its cell progeny included some with the translocated X and some with the normal X inactivated. Chimaeras with albino but not agouti, or agouti but not albino, patches would indicate either that X inactivation had already occurred by the time of injection; or that by chance cells of one donor type only had been selected in the ancestry of the melanocyte population; or, in the case of albino patches only, that the injected cell was genetically male (Fig. 14).

When donor cells were taken from blastocysts $3\frac{1}{2}$ days *p.c.*, nine of thirty-two young that survived to weaning showed chimaerism, with hairs of donor colour distributed widely throughout the coat despite the origin of their melanocytes from a single injected cell. Five of the nine chimaeras showed agouti patches, indicating that the injected cell had been female. Since four showed white patches also, it seems clear that X-inactivation, defined as an irreversible loss of gene function, has not taken place by $3\frac{1}{2}$ days *p.c.* A similar result was obtained using $4\frac{1}{2}$-day donor blastocysts (Gardner, 1974).

One conclusion of general validity for early development can be derived from all the studies that have been carried out on both injection and aggregation chimaeras. No instance has been found of a tissue or organ, however small, which in a chimaeric animal is always formed of one component only. This implies that it is groups of cells, rather than single cells, which are directed towards particular developmental pathways and thus give rise to particular tissues and organs. The consequences of this type of analysis will be explored in more detail in Chapter 10.

4

Sexual development

XX/XY chimaeras

Neither at the 8-cell nor at the blastocyst stage is it yet possible to sex a live mouse embryo. Chimaeras are therefore made without regard to the sex of the components, and since pre-implantation mouse embryos comprise about equal numbers of males and females (Vickers, 1967; Kaufman, 1973), 50 % of chimaeras on average would be expected to be of the XX/XY type.

The available information, such as it is, agrees well with this expectation. The four bodies of data summarized in Table 3 are consistent with one another, and give an overall proportion of XX/XY animals very close to the predicted 50 %. The paucity of the data reflects the tedium of chromosomal sexing in the mouse.

No *a priori* reasons exist for expecting XX/XY mice to be male, or female, or intermediate. In the samples listed in Table 3, Mystkowska & Tarkowski's nine XX/XY animals comprised two females (both foetal) and seven males (including one male hermaphrodite); the four of Milet *et al.* included a sterile female, two males and a female hermaphrodite; our eight were all males (six fertile and one sterile male, and one male hermaphrodite); and the four of Ford *et al.* consisted of two females and two males. A strong tendency for development in the male direction is evident, but a calculation of the overall proportion of males (76 %) may be misleading, as it could conceal a real heterogeneity between the different samples.

Mintz (1969a) mentions nineteen XX/XY chimaeras. These included one hermaphrodite, three phenotypic females (two sterile, one not mated) and fifteen males (eight fertile, five sterile, two not mated), but since they were stated not to be a random sample of the population, they shed little light on the relative frequency of the different phenotypes.

Of three overt chimaeras made in sheep by the injection technique, all were male (Tucker *et al.*, 1974). Two proved to be XX/XY, and one XY/XY.

Sex ratio

Indirect but more extensive data come from a consideration of the sex ratio of chimaeras. In the initial series of aggregation chimaeras described by Tarkowski (1961, 1963) a striking preponderance of males was seen (11/13),

TABLE 3. *The results of determining the chromosomal sex of aggregation chimaeras*

Author	XX/XX	XY/XY	XX/XY	XX/XY — total (%)
Mystkowska & Tarkowski (1968, 1970)[a]	3	3	9	60.0
Milet *et al.* (1972)[a]		4	4	50.0
McLaren (1975*a*)[b]	3	5	8	50.0
Ford *et al.* (1975*a*)[b]	2	4	4	40.0
Total		24	25	51.0

[a] Since the two component strains were mostly not distinguished chromosomally, the apparently non-XX/XY animals may have contained an undetected cell line of the other type; the proportion of XX/XY animals therefore represents a minimum estimate.

[b] The component strains were distinguished chromosomally; only those individuals in which the sex chromosome constitution of both cell lines was determined have been included.

together with three hermaphrodites. He interpreted this to mean that XX/XY individuals tended to develop as phenotypic males. Later evidence has upheld this interpretation, though it seems that the strength of the tendency varies from one strain combination to another.

The main bodies of chimaeric sex ratio data available in the literature are summarized in Tables 4 and 5 (modified from McLaren, 1972*a*). Where genetic markers are available (e.g. for coat colour) it is important to distinguish between animals which show evidence of chimaerism ('mixed-type') and those which fail to show such evidence ('single-type') and may therefore have lost one component entirely, since in the latter case there is no reason to expect the sex ratio to differ from that seen in a control, non-chimaeric population. In Table 4, the distinction between 'mixed-type' and 'single-type' animals is made, while Table 5 includes those bodies of data where no such distinction is possible.

In Table 4, the proportion of males among 'mixed-type' animals is well in excess of 50 % for every series except the so-called 'unbalanced' strain combinations of Mullen & Whitten (1971). In these combinations, one of the two component cell populations is represented only to a very limited extent in the coat, and perhaps also in the gonads. XX/XY animals would thus have gonads that were predominantly either XX or XY and might be expected to develop as females or males accordingly. The imbalance in these strain combinations is manifest also in the very low incidence of overt chimaerism (37.9 %). The data of Mintz (1969*a*), for those strain combinations in which

TABLE 4. *The sex ratio in those bodies of data where overt chimaeras ('mixed type', usually multi-coloured) are distinguished from 'single-type' animals*

Author	No. of strain combina-tions[a]	Overt chimaeras as % total mice from aggregated embryos	'Single-type'			'Mixed-type'			
			♂	♀	$\frac{\text{♂}}{\text{♂}+\text{♀}}$ (%)	♂	♀	⚥	$\frac{\text{♂}}{\text{♂}+\text{♀}}$ (%)
McLaren & Bowman (1969 and unpublished)	1	79.0	4	2	66.7	19	4	1	86.4[b]
Mystkowska & Tarkowski (1968)	1	77.8	1	1	50.0	8	1	1	88.9
Mintz (1969a)	2	43.3	71	103	40.8	71	48	4	61.0[b,c]
Mullen & Whitten (1971)									
'Balanced'	3	67.1	30	35	46.2	70	31	2	69.3[b]
'Unbalanced'	2	37.9	32	35	47.8	19	21	0	47.5[c]

[a] Where data from more than one strain combination have been combined, the sex ratios for the individual combinations were in every case homogeneous.
[b] ♂:♀ ratio significantly different from 1:1 ($P < 0.02$–0.001).
[c] ♂:♀ ratio significantly different from 3:1 ($P < 0.001$).

TABLE 5. *The sex ratio in those bodies of data in which it is not possible to distinguish overt chimaerism*

Author	♂	♀	⚥	$\frac{\text{♂}}{\text{♂}+\text{♀}}$ (%)	Significance level for deviation from 1:1
Tarkowski (1961)	11	2	3	84.6	$P < 0.01$
McLaren et al. (1972)	36	16	1	69.8	$P < 0.01$
Tarkowski (1969); Mystkowska & Tarkowski (1968, 1970)	27	18	1	60.9	$0.2 > P > 0.1$
Mintz (1969a)[a]	250	194	5	56.3	$P < 0.01$

[a] These data are derived by subtracting the values given in Mintz' Table 2 (two-colour and one-colour mice distinguished, see Table 4 above) from the overall values given in her Table 1 (Mintz, 1969a).

two-colour and one-colour animals are distinguished, show a significant excess of males in the two-colour (mixed-type) group. The corresponding figures for the 'single-type' animals show an excess of females, presumably a characteristic of the control population, so that if the two bodies of data are pooled, no excess of males is apparent. In the two-colour animals the excess of males does not amount to the 75 % expected if all XX/XY animals develop as males, but indicates that about 30 % are likely to be phenotypic females. Indeed, Mintz (1969a) mentions that at least three XX/XY females have been detected in her material. The relatively low incidence of overt chimaerism (43.3 %) suggests that these strain combinations may resemble in other ways too the 'unbalanced' combinations of Mullen & Whitten.

The sex ratio for those published series in which overt chimaeras are not identified is given in Table 5. Since 'mixed-type' and 'single-type' animals are confounded, the proportion of males would be expected to be less than 75 % even if all XX/XY individuals developed as males. It is clear that these data too lean in the male direction.

Determination of phenotypic sex

We may conclude that XX/XY individuals show varying degrees of dominance of the male component, ranging from 100 % development as phenotypic males in some strain combinations, to equal numbers of males and females in others. The variation seems to reflect the relative representation of XY and XX cells in the gonadal primordia, since it is 'unbalanced' combinations that have shown equality of the sexes. There is as yet no evidence of variation either in the specific masculinizing capacity of the Y chromosome of different strains, or in the response of XX cells. Dominance of the XX component has never been reported.

Hermaphroditism is rare, and where it occurs is often of the male type, with male organs (which may even be functional) on one side of the body and rudimentary female organs on the other. Tarkowski (1961, 1963, 1964a) described three intersexes among his newborn chimaeras, one with an ovotestis both sides, one with an ovary on the right side and an ovotestis on the left, and the third with an ovary on the left side and a testis on the right. Other bodies of data show a lower incidence of hermaphroditism, as listed in Tables 4 and 5. The hermaphrodite described by Mystkowska & Tarkowski (1968) showed an overwhelming predominance of XX cells in the bone marrow (96%), yet only the right side of the reproductive tract was bisexual, while the left was typically male. In Mintz (1965a) a phenotypically similar animal is mentioned, with male ducts and glands on both sides, a normal left testis with complete spermatogenesis, and on the right side an ovotestis accompanied by oviduct and uterus. The examples of Mystkowska & Tarkowski (1970) and McLaren et al. (1972) were both male foetuses, with one or both testes showing atypical regions lacking sex cords, with most of the germ cells in meiotic prophase; these gonads were interpreted as ovotestes. The hermaphrodite of McLaren (1975a) was an adult male who sired a litter of four at the age of 7 weeks, but subsequently proved sterile. When killed at the age of 20 months, the right side of the body showed male organs and an atrophied testis, while the left side showed a rudimentary uterus and ovary.

In human hermaphrodites with different degrees of masculinization of the two gonads (e.g. an ovary and an ovotestis, or an ovotestis and a testis), the right gonad is usually more masculinized than the left (Polani, 1970). This asymmetry is not evident in the chimaeric mouse hermaphrodites studied up to now.

What determines the phenotypic sex of an individual? The occasional differences between sides in hermaphrodites suggest that the determination is local, within the reproductive tract, rather than systemic, derived from elsewhere in the body. Secondary sexual characters develop in response to hormones produced by the gonads, and in particular the presence of a primitive testis is known to cause regression of the Müllerian duct derivatives (female) and development of the Wolffian duct derivatives (male). Hence the sex of the foetal gonads is of paramount importance, and this is likely to depend on the relative numbers of XX and XY cells within each during some critical phase of development. But a gonad contains two populations of cells of independent origin, the somatic component and the germ cells. Which population is responsible for deciding its sex? Arguments in both directions have been reviewed by Tarkowski (1969, 1970a), Short (1972) and Ford (1970).

In favour of the somatic component is the observation that in human XO/XY mosaics, the more masculinized gonad is usually characterized by a higher proportion of XY cells in cultures grown from its somatic tissue (Ford, 1970), and that the genital ridge undergoes its initial differentiation into a male or female gonad even if, for genetical reasons, germ cells are largely or perhaps entirely absent (Coulombre & Russell, 1954). Against the somatic component is the report by Mintz (1969a), confirmed by us, of fertile male XX/XY chimaeras with functional germ cells all XY yet with up to 95 % of the somatic tissue of the gonad made up of the XX cell population, as judged by enzyme determinations; and conversely, the finding by Ford et al. (1974) of a fertile female XX/XY chimaera with a great excess (98 %) of XY cells in the follicle cells of the ovary as judged by chromosome studies. However, all this evidence is concerned with adult gonads, and the somatic constitution of the adult gonad (particularly that of the follicle cells, which undergo extensive proliferation in postnatal life) is not necessarily a reliable guide to the constitution of the gonadal ridge in embryogenesis. The concordance between adult and embryo should be greater in the mosaic situation cited by Ford (1970), in which the two cell lines differ only in respect of a Y chromosome, than in the chimaeras of Mintz (1969a) and Ford *et al.* (1974), where strong selective forces are known to be operating.

Ford *et al.* (1974) find a strong correlation between right and left ovaries in the relative proportions of the two chimaera components among dividing follicle cells. They argue that the follicle cells in the two ovaries must therefore have a common origin. If these cells, contrary to common belief, enter the gonadal ridges secondarily, by migration, they need have no influence on sex determination and the 98 % of XY follicle cells in the XX/XY ovary requires no further explanation. If, as seems more likely, the left–right correlation implies that the gonadal ridges themselves have a common (presumably dorsal mid-line) origin, the 98 % of XY cells would indicate either that gonadal sex is not determined by the chromosome constitution of the somatic cells or that, as suggested above, strong differential proliferation has taken place between the embryo and the adult, perhaps during follicular maturation.

Milet *et al.* (1972) examined four XX/XY chimaeras, and found no relation between phenotypic sex and the proportion of XX cells in skin, spleen, kidney, bone marrow or cornea. They postulate that the sex of the gonad is determined jointly by the somatic tissue and the primordial germ cells, but they bring forward no direct evidence in support of their hypothesis.

In conclusion, the finding in experimental mouse chimaeras of XX/XY constitution that the somatic tissue of the adult gonad may be largely composed of cells discordant with its phenotypic sex does not necessarily contradict the widely held opinion that the sex of the gonad is determined by its somatic component, since it may be that the cellular make-up of the adult gonads bears little relation to that of the genital ridges in the embryo.

Freemartinism

The determination of phenotypic sex can be affected also by secondary chimaerism. In cattle twins, which nearly always share a common placental circulation, the female member of an opposite-sexed twin pair develops as a sterile freemartin. The fact that such sterile females are born co-twin to a male calf was first noted by John Hunter in the eighteenth century. The external genitalia of the freemartin are female, but uterus and oviducts are poorly developed or even absent and some male secondary sex organs (seminal vesicles, vasa deferentia, epididymides) are present. The gonads are small and usually contain testicular tissue; this is populated by germ cells during foetal life, but few if any germ cells survive beyond birth. The male partner is also affected, since bulls born co-twin to freemartins often become more or less sterile in later life. For a review of the consequences of secondary chimaerism for sexual development, see Short (1970).

Obviously something passes from one twin to the other to affect development, but what? Early workers assumed that the masculinizing influence on the female twin was exerted by a hormone: this is probably still the most plausible theory, though repeated failures to achieve masculinization by the experimental administration of testosterone and related androgens suggest that some as yet unidentified hormone may be responsible. The discovery of secondary blood chimaerism led to the suggestion that the modification of phenotypic sex might result from the presence in the female twin of cells with a male chromosome constitution. However, it is hard to envisage XY blood cells exerting a masculinizing effect on the gonad; moreover, in a series of heterosexual cattle twin pairs examined during foetal life the degree of masculinization of the female proved in no way correlated with the proportion of XY cells present in her liver (Vigier, Prepin & Jost, 1972). Ohno *et al.* (1962) suggested that primordial germ cells as well as blood cells were exchanged between cattle twins, but the observations have yet to be conclusively confirmed. Extensive progeny-testing of like-sexed cattle twins, as well as of bull calves born twin to freemartins, have failed to reveal any evidence of germ cell exchange. Thus the phenomenon of freemartinism, interesting though it is, has as yet shed no new light on the question of whether the phenotypic sex of a gonad derives from its germinal or its somatic elements.

In marmosets, almost all pregnancies end in twin births, and blood chimaerism occurs regularly, yet heterosexual twin pairs suffer no abnormalities of sexual development. Vascular anastomosis between the two placental circulations develops at a very early stage of pregnancy, so the failure to develop freemartinism is hard to understand. In Man too, opposite-sexed twins with XX/XY blood chimaerism (e.g. Miss Wa. and her brother, see Chapter 1) show normal sexual development.

Secondary sex glands

A striking example of cellular non-autonomy in the development of the reproductive tract was described by Mintz (1969*a*) and Mintz, Domon, Hungerford & Morrow (1972). In a sterile XX/XY male chimaera of the strain combination CBA-T6T6 ↔ C57BL, electrophoretic analysis of IDH isozymes showed that not only the testis, but also the epididymis and seminal vesicles, appeared to consist entirely of the XX (= CBA) component. The technique would have recognized a minor component making up more than 5 % of the tissue. This suggests that XX cells in the foetus, despite their genotype, were capable of responding to an androgenic stimulus by forming morphologically normal seminal vesicles.

The analysis was extended from the morphological to the biochemical level. Seminal vesicle fluid contains characteristic proteins not found in any female tissues. Electrophoretic analysis has revealed that these proteins differ from one inbred strain to another. Mintz *et al.* (1972) identified an

XX/XY male pseudohermaphrodite of constitution C3H↔C57BL, with a small penis and a vagina ending blindly. The seminal vesicle fluid from this animal contained seminal vesicle proteins characteristic of both component strains, proving first that XX cells can secrete a specifically male protein in response to a male environment, and secondly that the form of the protein is determined by the genotype of the XX cells themselves.

Germ cell differentiation in XX/XY chimaeras

In Amphibia, reciprocal grafts have shown that male germ cells developing in an ovary undergo oogenesis, and female germ cells in a testis undergo spermatogenesis (Blackler & Fischberg, 1961; Blackler, 1962). Whether mammalian germ cells could experience a similar degree of functional sex reversal was an open question until 1968.

If XX germ cells give rise to functional spermatozoa in XX/XY chimaeric males, the sex ratio of the progeny should be markedly distorted in a female direction. This distortion should be apparent in any large enough sample of the progeny of chimaeras, whether or not their sex chromosome constitution is known. Mintz (1968) obtained 29 384 young from 288 chimaeras: since the sex ratio of the progeny of individual chimaeras did not deviate significantly from that observed in control matings, she concluded that the XX germ cells did not undergo functional sex reversal. A direct test was provided by the combined progeny-testing and cytological analysis carried out by Mystkowska & Tarkowski (1968) on chimaeras in which one component carried the T6 chromosome marker. Two XY/XY males and an XX/XX female produced progeny of two genetic types, showing that both components of the chimaera had given rise to functional germ cells; on the other hand, two XX/XY males produced progeny of one type only, corresponding in both cases to the genotype of the XY component. Progeny-testing of a further sample of seven XX/XY males (McLaren, 1975a) in which the strain origin of each component was known, established that only one class of spermatozoa was produced, corresponding invariably to the XY component (Fig. 15).

Evidence from chimaeras thus makes it most unlikely that XX germ cells can be induced to undergo spermatogenesis in mice. Other lines of evidence, reviewed by Short (1972), lead to a similar conclusion for other mammals. Although it has been claimed that XX germ cells may enter the foetal testis in secondary chimaeras from heterosexual twin pregnancies in cattle and marmosets (see Tarkowski, 1969, 1970b), there is no evidence that such cells complete meiosis. In mice carrying the *sex-reversal* (*Sxr*) gene, which transforms genetic females into phenotypic males, all XX germ cells degenerate, but on an XO background active spermatogenesis occurs, though the spermatozoa produced are abnormal (Cattanach, Pollard & Hawkes,

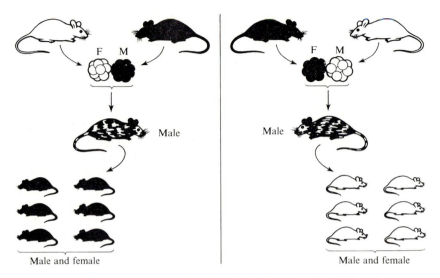

Fig. 15. Diagram of the breeding behaviour of XX↔XY male
chimaeras. F, female; M, male.

1971). The occasional 'fertile' testis tubule in XX *Sxr*/+ animals proved to
contain XO germ cells, arising presumably by non-disjunction (B. M. Cattanach & H. J. Mosley, personal communication). The fact that XO but not
XX germ cells can differentiate into spermatozoa suggests that the second
X chromosome is critical in preventing an appropriate response of the germ
cell to its testicular environment.

Little is known of the fate of XY germ cells in a chimaeric ovary, since only
two fertile XX/XY females have been reported (Ford *et al.*, 1975*a*). Of
twenty-three progeny, twenty-two proved to be of the genotype corresponding
to the XX component of the chimaera. The remaining animal, a male, could
have been derived only from the XY component. It proved to be of the rare
XXY chromosome constitution, with the Y chromosome derived from the
XX/XY mother, raising the intriguing possibility that an XXY germ cell
arose by non-disjunction from the XY cell line, and that the presence of a
second X chromosome, which in the testis militates against spermatogenesis,
enabled the germ cell in the ovary to undergo oogenesis. Alternatively, an
XY germ cell may have undergone oogenesis, with non-disjunction at the
first meiotic division. To what extent either XY or XXY germ cells normally
undergo oogenesis in XX/XY females remains an open question, since as
Ford *et al.* (1975*a*) point out, the apparently undisturbed sex ratio of the
progeny (Mintz, 1968) is much less conclusive evidence against functional
sex reversal in female chimaeras than it is in males.

XX germ cells in XX/XY males not only fail to form spermatozoa, but are not even to be found among primary spermatocytes (Mystkowska & Tarkowski, 1968). A possible clue to their fate came from the observation by Mystkowska & Tarkowski (1970) that the testes of male XX/XY chimaeras on the 16th to 17th day of foetal development contained germ cells in meiotic prophase, mostly in pachytene. In the normal female mouse the germ cells enter meiosis before birth, on the 14th or 15th day of gestation. In the normal male, on the other hand, the germ cells do not enter meiosis until several days after birth, so that the foetal testis contains large numbers of spermatogonia undergoing mitosis, but no meiotic stages. In a later study, germ cells in the early stages of meiosis were found in the testes of foetal chimaeras up to $17\frac{1}{2}$ days *p.c.* (Fig. 16), but they failed to reach diakinesis and by $18\frac{1}{2}$ days they were degenerating (McLaren *et al.*, 1972). Many oocytes in the normal ovary degenerate at a corresponding stage of foetal development.

The obvious explanation of these meiotic cells is that they are XX germ cells, behaving autonomously and entering meiosis at the time laid down by their genotype. An alternative possibility (see Fig. 17) is that the somatic tissue of the gonad determines the time of entry of germ cells into meiosis, so that those germ cells of either sex in the XX/XY testis that happen to be surrounded by a large enough patch of XX somatic tissue are driven into meiosis, while the remainder (XX and XY) continue to divide mitotically under the influence of XY somatic tissue. This explanation, favoured by Tarkowski (1969), is more easily reconciled with the low proportion (about 5 %) of germ cells observed in meiosis in the XX/XY testis: although XX and XY germ cells are presumably present in more or less equal proportions, the effect of XX somatic tissue could involve a threshold phenomenon such that only the occasional patch of unmixed XX cells was effective.

The possibility that the cells in meiosis include XY germ cells is not supported by the observations of McLaren *et al.* (1972). Injection of pregnant females with tritiated thymidine showed that the meiotic germ cells in chimaeric testes replicated their DNA on the same day as did those in normal foetal ovaries, and developed at the same rate. No sign was seen of the striking late-labelling pattern characteristic of meiotic prophase in male germ cells (Kofman-Alfaro & Chandley, 1970), nor of the prominent sex vesicle that contains the XY bivalent. It therefore seems that the cells in meiosis are XX rather than XY germ cells. Their relative scarcity, and the differences reported in their incidence in the two testes of the same foetus, might suggest that some influence of XX somatic tissue is also required if meiosis is to begin (i.e. a combination of the two hypotheses illustrated in Fig. 17).

The degeneration of meiotic germ cells before birth in the mouse is presumably due to the adverse hormonal environment of the testis, and the absence of the layer of follicle cells that normally surrounds the oocyte at this time. Some may survive into the postnatal period, since Mystkowska & Tarkowski (1968) found growing oocytes in one area of the testis of a male

4

(a)

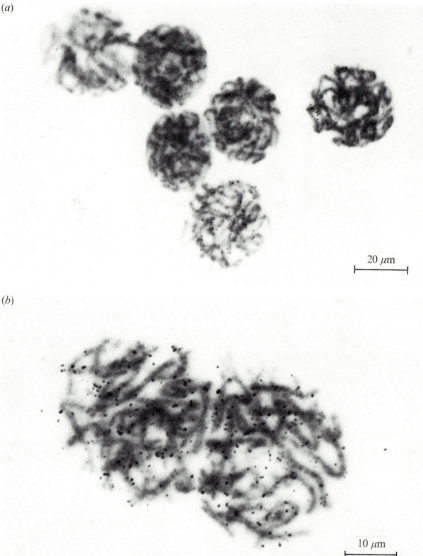

20 μm

(b)

10 μm

Fig. 16 (a) and (b). Autoradiographs of air-dried germ cells, labelled with tritiated thymidine, in the early stages of meiosis from the testis of a foetal chimaera 16½ days *p.c.* No sex vesicle or XY bivalent is apparent. From McLaren, Chandley & Kofman-Alfaro (1972).

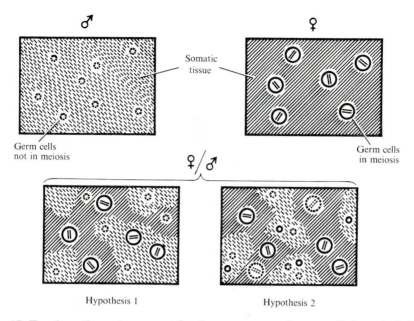

Fig. 17. Two hypotheses to account for the appearance of germ cells in meiosis in XX↔XY chimaeras before birth. On hypothesis 1, female germ cells enter meiosis regardless of their cellular environment; on hypothesis 2, all germ cells surrounded by XX somatic tissue enter meiosis, whether they themselves are XX or XY in constitution. Somatic tissue is indicated by cross-hatching, broken for male, continuous for female. Germ cells are represented by circles, broken for male, continuous for female, small if not in meiosis, large if in meiosis. From McLaren (1972*a*).

chimaera as late as 5 days after birth, but it seems most unlikely that XX germ cells are able to complete even the first meiotic division in the testis of any mammal.

Progeny of single-sex chimaeras

When both components of a chimaera are of the same chromosomal sex, XX or XY, both germ cell populations can form functional gametes and mixed progenies may result. Mixed progenies were first reported by Mystkowska & Tarkowski (1968) and Mintz (1968). To discover what proportion of chimaeras produce both types of gamete, attention must be restricted to overt chimaeras, in which two populations of cells are known to co-exist.

The available data on overt chimaeras are summarized in Table 6. The situation is complicated by the fact that a proportion of the animals in each sample will be XX/XY chimaeras, and hence most unlikely to produce mixed progenies. The proportion of XX/XY individuals is unknown in most samples,

TABLE 6. *The breeding performance of overt chimaeras*

Source	Sex	Total sample			XX/XX and XY/XY only	
		Mixed pro- genies	Single pro- genies	% mixed	Mixed pro- genies	Single pro- genies
Mystkowska & Tarkowski (1968)	♂	2	3	40.0	2	1
Mintz (1969*a*)	♂	9	14	39.1	—	—
Mullen & Whitten	♂	11	35	23.9	—	—
(1971) 'balanced'	♀	13	6	68.4	—	—
Mullen & Whitten	♂	1	16	5.9	—	—
(1971) 'unbalanced'	♀	7	10	41.2	—	—
McLaren (1975*a*)	♂	1	15	6.3	1	4
	♀	0	4	0.0	0	3
Ford *et al.* (1975*a*)	♂	3	5	37.5	2	2
	♀	5	3	62.5	2	0

but is expected to be about 50 % (see Table 3), and may be up to two-thirds of male chimaeras. The single-progeny animals of Mystkowska & Tarkowski (1968), Mintz (1969*a*) and Ford *et al.* (1975*a*) could therefore well be all or mostly XX/XY chimaeras: indeed, an XX/XY chromosome constitution was identified by Mystkowska & Tarkowski in two out of three animals, and by Ford *et al.* in three out of seven whose chromosomes were studied. Mullen & Whitten (1971) report significantly more than 50 % 'single-progeny' chimaeras in their 'balanced' combinations, and a large excess of single progenies in the 'unbalanced' ones (26/34). In the data of McLaren (1975*a*) the excess was even greater (19/20), and at least seven of the 'single-progeny' animals were shown not to be XX/XY chimaeras. Neither type of germ cell seemed at a selective advantage, nor did the strain combination appear markedly 'unbalanced' in the other tissues examined.

It seems therefore that, at least in some strain combinations, one or other cell line is often by chance excluded from the germ cell population. This suggests that the two cell populations in the chimaeric embryo may not be randomly mixed at the time when the germ cells are set aside.

The distribution of the two populations of functional germ cells within the gonads of those individuals in which they co-exist has not yet been investigated. Indirect evidence, based on the proportions of the two types of progeny in successive litters, suggests that in some strain combinations the distribution is far from random. E. T. Mystkowska (personal communication) reports two large mixed progenies from CBA-T6↔CBA-p male chimaeras,

TABLE 7. *Segregation of progeny in 12 litters sired by a male chimaera paired with 6 females (a–f) of the recessive stock*

Date of birth of litter	Female	Number of progeny	
		'Recessive'	'Dominant'
21 Apr. 1972	a	0	6
12 June 1972	a	4	0
16 Oct. 1972	b	7	0
6 Nov. 1972	b	5	1
20 Dec. 1972	c	5	1
25 Dec. 1972	d	5	1
21 Jan. 1973	e	7	0
8 Mar. 1973	b	0	6
9 Mar. 1973	c	6	0
22 Mar. 1973	d	1	1
29 Apr. 1973	b	8	0
5 May 1973	f	0	3

From McLaren (1975*a*).

in which the proportions of the two types of young do not vary significantly between litters ($\chi^2_{(28)} = 27.0$; $\chi^2_{(26)} = 29.0$); on the other hand, McLaren (1975*a*) found that the proportions of the two types of young sired by male chimaeras involving a multiple recessive strain shifted very significantly from one litter to another throughout reproductive life (e.g. Table 7). The variations did not seem to follow any regular trend: they were not related to the age of the male, nor to the age or identity of the female bearing the litter. Perhaps the area of a spermatogenic tubule that is colonized by the descendants of a single germ cell or patch of germ cells can be extensive enough to occupy most of an ejaculate; perhaps indeed this area comprises an entire tubule, with the tubules being emptied one at a time into the common collecting duct. In chimaeras involving the multiple recessive strain, patches of the two types of cell may be relatively large. The absence of any heterogeneity in the CBA-T6↔CBA-p progenies may reflect the close genetic relationship between the components in this strain combination.

The lack of an age effect is in contrast to the situation described by Mintz (1968, 1969*a*). When C57BL↔C3H/He chimaeras were mated to C57BL partners, the proportion of C57BL young decreased sharply with time when male chimaeras were used. When the chimaeric partner was the female, although C3H germ cells still appeared to predominate, there was said to be no change with time. The effect in males was interpreted as indicating continuous selection against C57BL germ cells in the chimaeric testis; in the female the germ cell population is established prenatally, and no further cell proliferation occurs, so no change with time would be expected. A possible alternative explanation of the apparent male effect is suggested by the preliminary studies of Burgoyne (1973). Two C57BL↔C3H/Bi males again showed a predominance of C3H germ cells, but the proportion of C57BL young among the progeny was related to the parity of their C57BL female partners. The proportion was highest among second and third litters, and was very low in later litters. One male still showed 35 % of spermatozoa of the C57BL type when killed at 15 months. This suggests that intra-uterine selection might be acting against the inbred C57BL embryos, the intensity of the selection varying with the age and genotype of the mother so that inbred females, as they grew older, proved less able to support development of the inbred embryos.

The preponderance of C3H germ cells in C3H↔C57BL chimaeras is particularly striking, since in the somatic tissue of the testis the selection is in the reverse direction, with C57BL tissue tending to predominate (Mintz, 1969a). No relation was found between the strain composition of the germ cells and that of the rest of the testis: indeed, in some males the germ cells were all from one component, and the somatic tissue largely from the other.

The tendency for one component to predominate in the composition of the chimaeric germ cell population, even in 'balanced' strain combinations, is seen in combinations other than C3H↔C57BL, for example SJL↔C57BL, in which most of the males produced only SJL spermatozoa and even those producing both types showed a marked preponderance of SJL (Mullen & Whitten, 1971).

When the two components of a chimaera differ at more than one genetic locus, a further test for interaction between gametes and somatic tissue becomes possible. It is widely assumed that the genetic content of a gamete is wholly unaffected by the cellular environment in which it develops. This assumption is based on little experimental evidence: since it is usual for germ cells (at least until meiosis) and somatic tissue to share a common genetic constitution, any interchange of genetic material would pass unnoticed. In a chimaera, as we have seen, the germ cells may differ genetically from the surrounding gonadal cells, yet even here the transfer of genetic information from somatic tissue to germ cells would be hard to detect so long as only one scorable genetic difference distinguished the two components. Such a transfer would at most appear as an increased number of germ cells of one genetic type: indeed, the increase with time of progeny sired by the C3H component in C3H↔C57BL males (Mintz, 1968) could be interpreted in this way.

A stringent test for the occurrence of genetic transfer is provided by a series of chimaeras described by McLaren (1975b), in which the two components differed at ten loci. Chimaeric males were back-crossed to females of the homozygous recessive component strain, and 1851 progeny were born, of which 1092 were from germ cells derived from the multiple recessive component. The progeny segregated cleanly into two types, corresponding to the two chimaera components: no intermediates were observed, and no 'recombinant' phenotypes. Any somatic contamination of the genetic content of the germ cells, at least for these ten loci, must therefore be exceedingly rare, if it occurs at all.

Sperm phenotype

The shape of spermatozoa varies significantly from one inbred strain of mice to another (see Beatty, 1972). This genetically determined feature could be determined by the genotype of the germ cell itself: since F_1 hybrid males show no greater variation in sperm phenotype than do inbred males, the genetic information would need to be transcribed during the diploid phase of spermatogenesis, unless the intercellular bridges that link 'brother' spermatids allow diffusion of gene products between post-segregational, haploid cells. Alternatively the genetic effect could operate through the cellular environment, and thus be dependent on the genotype of the somatic cells of the testis, for example the Sertoli cells which form intimate associations with the germ cells during spermatogenesis. In normal males there is no means of deciding between these two possibilities. In chimaeras, on the other hand, there are two genetically distinct populations of somatic cells in the testis, so that a germ cell of one genotype may well develop in association with a Sertoli cell of contrasting genotype. If sperm phenotype were affected by the cellular environment, a chimaera shown by breeding tests to be producing gametes from one component only might show spermatozoa resembling in phenotype those of the other strain, or alternatively some intermediate phenotypes might be produced.

This approach has been employed by Burgoyne (1975). He used the C57BL↔C3H/Bi strain combination, and made measurements of several sperm characteristics, including head length, head breadth and mid-piece length. Using discriminant functions, it proved possible to distinguish between the spermatozoa of the component strains with an accuracy of 2 %, in the sense that there were only two chances in a hundred that the strain of origin of a spermatozoon would be incorrectly identified.

Minor but significant differences were observed between the dimensions of spermatozoa from chimaeras and those from control males. These did not indicate a response, even a minor one, to the chimaeric nature of the somatic tissue of the testis, since in males producing two types of spermatozoa the means for both populations tended to shift in the same direction, keeping the distance between the means unchanged. Further, the same shifts were observed in spermatozoa from aggregants that showed no chimaerism, suggesting that the explanation might lie in some maternal influence exerted by the randomly bred Q-strain foster-mothers.

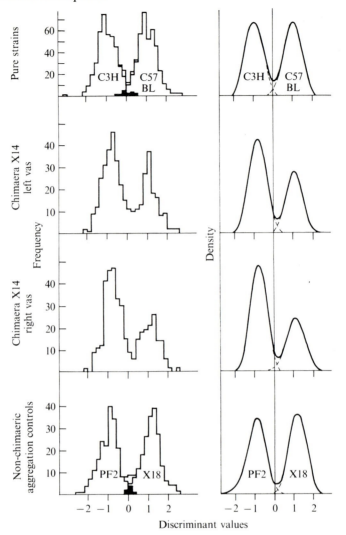

Fig. 18. The shape of spermatozoa from C3H and C57BL males, and from C3H↔C57BL chimaeras. Histograms are given on the left, and smoothed curves on the right, for discriminant values based on head length, head breadth, head area and mid-piece length. The non-chimaeric aggregation controls included one of C3H type (PF2) and one of C57BL type (X18). The chimaera, X14, produced two types of progeny, one from C3H and one from C57BL spermatozoa. From Burgoyne (1973).

When allowance was made for this minor shift in dimensions the results were unequivocal. Chimaeras produced spermatozoa closely resembling in their phenotypes the component strains, with no intermediate forms. Males siring mixed progenies showed two types of spermatozoa (Fig. 18), while those producing progeny of one type only showed a single population of spermatozoa of appropriate phenotype. The genetic differences in sperm phenotype were clearly controlled by genes acting in the germ cells themselves.

5

Pigment patterns

The value of chimaeras for the study of development, and in particular for the analysis of cell lineages, depends on the availability of suitable markers. One of the most satisfactory intrinsic markers available for this purpose is pigment (usually melanin). As pointed out in Chapter 2, two locations in which pigment remains within the melanocyte that produces it are the inner ear and the pigmented retina of the eye. Here pigment can act as a cell marker, enabling the distribution of the two chimaera components to be analysed at a cellular level. This is not so in hair or in the choroid layer of the eye, where pigment is secreted by melanocytes into extracellular spaces; but even here, pigment patterns in chimaeras may provide information on developmental processes and on gene action. Similar information can be derived from the analysis of females heterozygous for X-linked genes affecting pigment, owing to the phenomenon of X-inactivation. A comparison between chimaeras and X-inactivation mosaics will be made in Chapter 9.

Hair

For diagnosing chimaerism, coat colour offers a more obvious and easily visible system than any other character. The first adult mouse chimaeras to be made and identified were between an agouti strain and a non-agouti (black) strain. They were mottled, showing both agouti and black coloration arranged in 'a striking pattern of markings' (Mintz, 1965b).

In the mammalian coat, pigment is secreted into hairs by melanocytes, descendants of the melanoblasts that migrate down from the neural crest during embryonic development, to colonize the hair follicles. Thus a single hair follicle in a chimaeric individual may contain not only two genetically differing populations of epidermal cells, but also two independently derived populations of melanocytes. Coat colour can be affected by genes (e.g. *brown, albino, dilute*) acting through the melanocytes to determine the type or quantity of pigment synthesized, or the form in which it is laid down; it can also be affected by genes acting through the epidermal cells of the hair follicle. Transplantation experiments (Silvers & Russell, 1955) suggested that *non-agouti* acts through the hair follicle; this has been fully confirmed by chimaera studies. Cattanach, Wolfe & Lyon (1972) have suggested, on the basis

(a)   (b)   (c)

Fig. 19. Hair shafts from (*a*) an agouti hair (C3H × C57BL F$_1$), (*b*) a cream hair (multiple recessive, *dd bb aa pp c^{ch} c^{ch}*), (*c*) a chimaera. The septa in the chimaera hair vary in the amount of melanin pigment they contain, from none as in the multiple recessive, to the full amount of the agouti component.

of studies with chimaeras, that at least one gene at the *Mottled* locus may act independently through melanocytes and hair follicle cells. In general, however, genes affect one cell population or the other, but not both. As we shall see later, 'melanocyte chimaeras' present a very different appearance from 'hair follicle chimaeras' from the point of view of the coat colour pattern.

Microscopic examination has shown that chimaerism may extend to the level of the individual hairs (McLaren & Bowman, 1969). In chimaeras between a pigmented and a non-pigmented strain the shafts of some hairs were fully pigmented, others lacked pigment entirely, and others again were patchy, with adjacent septa differing widely in the amount of pigment they contained (Fig. 19). Various degrees of intermediate expression were also seen in the hairs of agouti↔non-agouti chimaeras, corresponding presumably to hair follicles with different proportions of the two component genotypes. The presence of chimaeric hairs in melanocyte chimaeras, confirmed by Mintz & Silvers (1970) and by Cattanach *et al.* (1972), establishes that more than one melanoblast may colonize a single hair follicle.

Variegation at more than one locus

When chimaerism involves more than one locus affecting coat colour, phenotypic interactions may arise in the formation of pigment. McLaren & Bowman (1969) combined embryos from a strain homozygous for five recessive genes affecting coat colour, with embryos carrying the corresponding dominant alleles. In appearance, the two component types were respectively cream-coloured and agouti ('wild-type'). Four of the loci (*dilute, brown, pink-eye, chinchilla*) affect the amount or type of pigment throughout the hair shaft, so that in combination they result in a cream hair lacking melanin entirely except for a short region near the base; the fifth, *non-agouti*, determines the type of pigment (black eumelanin rather than yellow phaeomelanin) laid down near the tip of the hair. Thus the hairs of an agouti mouse appear brown because of a yellow band located near the tip.

Thirty individuals of this strain combination (of which eleven were described in McLaren & Bowman, 1969) survived to the age of 2 weeks, when their coat colour was assessed. Four were all cream-coloured and two were all agouti; of the remaining twenty-four overtly chimaeric animals, six showed both cream and agouti coloration, one was cream and black, three were agouti and black, ten were tri-coloured (cream and agouti and black) and four were cream with some dark pigment. Microscopic examination confirmed the presence of black hairs.

The emergence of black hairs indicated that melanoblasts of the 'dominant' component, homozygous for wild-type alleles at the *dilute, brown, pink-eye* and *chinchilla* loci and therefore fully competent to synthesize pigment, had colonized hair follicles consisting predominantly of 'recessive' epidermal cells. At the *agouti* locus, the melanocytes would therefore carry the wild-type *agouti* allele and the hair follicle the recessive *non-agouti* allele; the hair was black, lacking the yellow band of the agouti, because the type of pigment synthesized by the melanocytes was determined not by their own genotype, but by the genotype of the hair follicle in which they functioned. The situation is illustrated in Table 8.

Using a different combination of strains, black agouti (*BB AA*) with brown non-agouti (*bb aa*), Mintz & Silvers (1970) produced chimaeras showing the two component colours plus two 'recombinant' colours, brown agouti (*bb AA*) and black non-agouti (*BB aa*), formed by melanoblasts colonizing hair follicles of contrasting genotype.

TABLE 8. *Hair colour, showing phenotypic interaction: the emergence of black hairs in chimaeras between cream ('recessive') and agouti ('dominant') mice*

Genotype of melanocytes	Genotype of hair follicle cells	
	'Recessive'	'Dominant'
'Recessive'	Cream	Cream
'Dominant'	Black	Agouti

Patterns in melanocyte chimaeras

The first description and attempted analysis of the coat colour patterns of melanocyte chimaeras was given by Mintz (1967a). In a number of different albino ↔ coloured strain combinations, she observed a widespread tendency for broad transverse bands of colour on the head, body and tail. On the antero-posterior borders of the bands, adjacent colours were mingled to varying degrees, and in some cases the brindling extended throughout the band. At the mid-dorsal line, on the other hand, there was a sharp discontinuity, so that the pattern on the two sides of the animal seemed to be independently determined.

Mintz interpreted each band as representing a clone of melanoblasts, descended mitotically from a single progenitor cell in the neural crest. The transverse orientation of the bands reflects the route of migration of the melanoblasts laterally from the neural crest, at 8–12 days of gestation. She pointed out that the independence of left and right sides would follow if the clones were initiated at a time when the neural crest cells could not pass from one side of the embryo to the other, that is before the longitudinal neural folds had come together and fused in the mid-line (Mintz, 1970b). She suggests 5–7 days of gestation as the most likely period of initiation of melanoblast clones. A similar appearance was seen when two positive colours were combined, such as black↔brown (bb), black↔dilute brown (bb dd), black↔ leaden (ln ln). Aggregations involving the black-eyed white locus (Mi^{wh}, mi^{bw}), in which melanocytes are absent rather than merely amelanotic, gave a lower incidence of melanocyte chimaerism than did those involving albino, but the white bands that occurred were particularly sharply defined (Mintz, 1971b).

On the basis of her initial observations, Mintz (1967a) recognized a 'standard pattern', which she claimed was the 'basic type, obtained most frequently', consisting of seventeen bands down each side of the animal, three on the head, six on the body, and eight on the tail. She inferred that there must be thirty-four primordial melanoblasts in the neural crest, a chain of seventeen on each side, with the two genetically different types (e.g. black and white) occupying 'alternating rather than random positions in the chains'. However, this seemed biologically implausible, and in later papers (Mintz, 1970*b*, 1971*b*) she withdrew the suggestion that the two types of melanoblast were arranged non-randomly. A random arrangement implies that adjacent melanoblasts would often be of the same colour, so that the number of visible bands would almost always be less than the number of 'primordial melanoblasts'. Indeed, thirty-four melanoblasts would only give rise to thirty-four bands in less than one in a million random arrangements. Wolpert & Gingell (1970) calculated that, if the phenotype 'obtained most frequently' showed seventeen bands each side, the most likely number of 'primordial melanoblasts', assuming equal proportions of the two types arranged randomly, would be about thirty-two per side, i.e. sixty-four in total. Using a slightly different mathematical approach, West (1975*a*) estimated that thirty-four clones each side would be required to produce seventeen bands.

In fact it is now clear (Mintz, 1970*b*, 1971*b*) that the 'standard pattern' denotes not the most frequent phenotype, but an abstraction or 'archetypal design' representing the greatest possible number of bands, i.e. 'the maximal possible clonal phenotypic differentiation, which is only seen in those cases where clonal colors happen to alternate, affording individual clonal identifications' (Mintz, 1970*b*). Thus the 'standard pattern' represents 'the developmental "signal" or undisturbed picture' that displays 'the clonal components of the melanoblast system', while other, so-called 'derived', patterns are 'modifications based on the single archetype, and comprise orderly pattern alternatives as well as developmental "noise"' (Mintz, 1971*b*). In 'developmental noise', Mintz includes such disturbing factors as cell mingling, cell death and differential rates of proliferation.

The concept that, in melanocyte chimaeras, each band of colour marks an area colonized by melanoblasts of one component type is supported by the skin-grafting experiments of Mintz & Silvers (1970). Chimaeras were made between strains carrying different melanocyte markers and also different alleles at the *H-2* histocompatibility locus. In skin grafts to the component strains, there was selective survival of host-type melanocytes, while cells of the other component strain were briskly rejected. Rejection was highly specific, even when both types of melanocyte co-existed in a single hair follicle, demonstrating that histocompatibility antigens are produced by melanoblasts themselves and do not move from one cell to another.

The transverse bands characteristic of melanocyte chimaeras may be interpreted as regions within which melanoblasts move freely, but between which little melanoblast mingling takes place in normal development. The number of such regions in the head, body and tail may vary little from strain to strain. The chimaeras described by McLaren & Bowman (1969) were

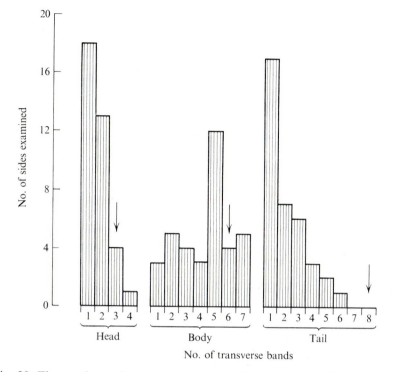

Fig. 20. The numbers of transverse bands or pigmentation 'regions' (mainly pigmented, mainly unpigmented, or mixed) in the head, body and tail on each side of eighteen melanocyte chimaeras of the strain combination described by McLaren & Bowman (1969). The arrows indicate the number of bands in the 'standard pattern' that Mintz regards as representing the maximum number to be expected on a clonal analysis.

examined from this point of view, the number of more or less discrete transverse territories recorded, and a note taken of whether they were predominantly light or dark in colour, or mixed. The numerical data are shown in Fig. 20: the maximum number of regions agrees well with the expectations of Mintz's 'standard pattern'. Errors of classification will contribute to the variability; some biological variation must also be expected.

Whether or not each region is necessarily colonized by the descendants of a single primordial melanoblast, as postulated by Mintz, seems more debatable. Lyon (1970, 1972) pointed out that such regions could be colonized by two or more melanoblasts; when these were of the same genotype a single-colour band would result and when they were of differing genotype the band would appear brindled. Certainly regions of mixed colour are common in melanocyte chimaeras, and show little consistent

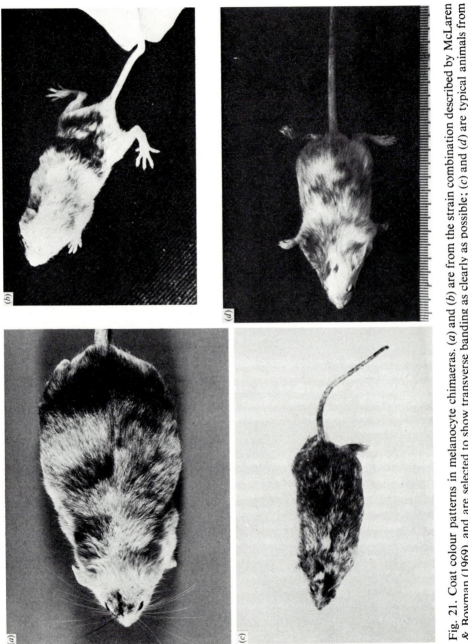

Fig. 21. Coat colour patterns in melanocyte chimaeras. (*a*) and (*b*) are from the strain combination described by McLaren & Bowman (1969), and are selected to show transverse banding as clearly as possible; (*c*) and (*d*) are typical animals from the series described by Mystkowska & Tarkowski (1968), showing much less clear-cut banding.

tendency for a single-colour core, as would be expected if they always arose through cell-mingling across the borders of adjacent regions.

Cattanach (1974) has used a spotting gene *s* to reduce melanocyte movement and hence make the coat colour pattern sharper, in an analysis of melanocyte patterns in *flecked* X-inactivation mosaics (see Chapter 9). His results are in general agreement with those of Mintz, showing three bands on the head and six or seven on the body. Each band seems to be colonized by a single clone, with the possible exception of the last body band, on the rump. Here the frequency of mixed bands is high, suggesting that colonization may involve more than one melanoblast clone.

The extent to which melanocyte chimaeras show a detectable pattern of transverse bands, rather than an overall brindled appearance, varies from one strain combination to another. Banding was apparent in all the strain combinations studied by Mintz (1967*a*), and also in that of McLaren & Bowman (1969) (Fig. 21*a* and *b*). On the other hand, CBA *pp*↔CBA *T6T6* chimaeras show less evidence of banding (Fig. 21*c* and *d*), either during life (Mystkowska & Tarkowski, 1968) or on examination of the pelts after death (McLaren, unpublished observations). In this combination the two components differ at only a single gene locus. It seems possible that a wider genetic divergence between the chimaera components might militate against random mixing, leading to a tendency for genetically similar cells to remain together within the neural crest, as suggested by McLaren (1969). This would favour single-colour rather than mixed bands. Even after the melanoblasts have left the neural crest, the regularity of the coat colour pattern must depend critically on the time at which clonal proliferation begins to dominate over cell mingling. The strain combination used by McLaren & Bowman (1969) showed a tendency for non-random assortment of the component cell populations (i.e. a large 'patch size') not only in the coat colour pattern but also in the retina (see below) and in the spermatogenic tubules (see Chapter 4). In the coat, single-colour transverse bands were often seen, and there was also a significant tendency for there to be more cream-coloured recessive-type hairs anteriorly, so that the anterior part of the animal appeared lighter than the posterior (West & McLaren, 1976; see also Chapter 10).

Patterns in hair follicle chimaeras

A hair follicle may affect the appearance of the hair growing within it, including its colour, either by modifying the activity of the melanocytes, as with genes at the *agouti* locus, or by altering the structure of the hair, as with genes such as *fuzzy* and *tabby*. Mintz (1970*b*) reported that mice chimaeric for *fuzzy* showed a 'standard pattern' which resembled that of melanocyte chimaeras in showing transverse bands with a mid-dorsal discontinuity, but differed in that the bands were much finer. A similar appearance of very fine transverse bands in the maximally patterned individuals was seen in agouti↔non-agouti chimaeras, together with a distinctive set of markings on the head (Mintz, 1971*b*; T. G. Wegmann, personal communication; McLaren, unpublished observations). Fine bands were also seen in chimaeras at the *tabby* locus (Cattanach *et al.*, 1972; Grüneberg, Cattanach, Wolfe, McLaren & Bowman, 1972), resembling the banding of heterozygous $Ta/+$ females (see Chapter 9).

If each hair follicle band is made up predominantly of cells of one of the two chimaera components, so that the colour of the coat in a given area directly reflects the genotype of the cells in that area, then the pattern of banding may reflect the clonal history of the tissue, as is thought to be the case for melanocyte chimaera patterns. Mintz (1969*b*, *c*; 1970*b*; 1971*b*) suggested that each narrow band represents a single hair follicle clone and pointed out that, in the region of the body where somites can be seen during embryonic development, the maximum number of narrow bands corresponds well with the number of somites. She therefore postulates that the hair follicle pattern reflects a mesodermal clonal pattern, of somite origin. The mesoderm is known to contribute to the hair follicle, in the form of the dermal papilla. The number of such clones is estimated to be 75–100 on each side, so that a single melanocyte area (whether containing one clone, as Mintz believes, or several) would span about six hair follicle clones.

An alternative interpretation would be that the hair follicle bands do not represent localized clones of contrasting genotype, but rather reflect some underlying threshold phenomenon superimposed on a systemic wave pattern which arises from the presence of two genetically distinct cell populations. The *agouti* locus is known to be strikingly responsive to positional information, in that hair follicles of identical genotype may produce yellow hairs on the belly, agouti hairs on the sides, and black hairs along the middle of the back. Indeed, the type of pigment synthesized by A^+ melanocytes in culture can be controlled by the concentration of sulphydryl compounds in the medium (Cleffman, 1963). Agouti patterns have been described as 'a visible expression of the metabolic equilibrium of the animal' (Wolff, 1971).

On the systemic 'wave' hypothesis neighbouring bands should be rather similar in width. On the 'clonal' hypothesis, assuming an initially random arrangement of two components present in equal amount, bands of unit width should be equal in number to double-width bands, with wider band widths represented but occurring progressively less often. No measurements or counts of bands on hair follicle chimaeras or on tabby heterozygotes have been published, but in both situations the banding appears rather regular, more so than in melanocyte chimaeras. Another observation, hard to explain on the clonal hypothesis, is that certain areas of the body appear to be almost always occupied by one type of hair follicle: for example, the pattern of agouti↔non-agouti chimaeras has a preferentially non-agouti area between

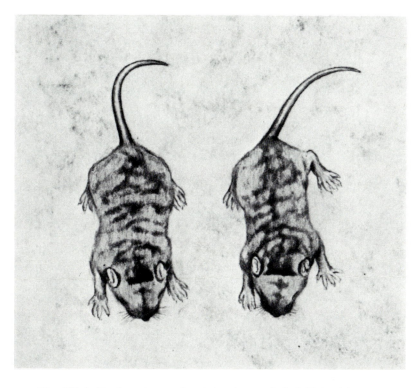

Fig. 22(*a*). Black and agouti striping in two hair-follicle chimaeras
(C3H↔C57BL).

the ears (Fig. 22*a*). This tendency for pigmentation to be distributed non-randomly within the pattern may be indicative of a systemic influence.

In an attempt to distinguish between the two possibilities, Mintz & Silvers (1970) made skin grafts from agouti↔non-agouti chimaeras onto the component inbred strains, which carried different *H-2* alleles. Unfortunately the results were less clear-cut than in the equivalent analysis of melanocyte chimaeras by skin-grafting. In melanocyte chimaeras single-colour areas are large, so that an entire graft could bear hairs all of one colour. When placed on a recipient of contrasting *H-2* genotype all donor melanocytes would then be rejected, leaving the hairs white. In hair follicle chimaeras, on the other hand, the narrowness of the bands implies that grafts will include both types of hair; partial rejection of the graft will take place, with elimination of all cells of the foreign *H-2* type, and on either hypothesis the new hairs that grow would be expected to be of recipient type. This result was obtained, and in addition some of the new hairs showed striking morphological abnormalities, presumably due to partial destruction of an initially composite hair follicle.

When chimaeras are made between strains differing at two alleles, both acting through the hair follicles, the relative distributions of the two phenotypes can be examined. Mintz & Kindred (cited by Mintz, 1970*b*) combined strains differing both at the *agouti* locus, and at the *fuzzy* locus that affects hair structure. The component strains were fuzzy non-agouti and normal agouti, but the chimaera also showed all intermediate types of hair structure and agouti banding, in various combinations. The appearance of intermediate types establishes that a hair follicle originates from more than one cell. The variety of combinations of hair structure and banding reported raises the suspicion that *fuzzy* and *agouti* may operate on two different cell populations within the hair follicle: no information was given on the relative incidence of the various combinations, nor on the incidence of 'component' (fuzzy non-agouti and normal agouti) versus 'recombinant' (fuzzy agouti and normal non-agouti) hairs.

In a similar experiment a strain homozygous for *waved-2* and *non-agouti* was combined with a straight-haired agouti strain (McLaren & Bowman, 1969). Agouti pigmentation is controlled by the dermal component of the hair follicle (Mayer & Fishbane, 1972); *waved-2* is thought to involve a lack of keratinization, due presumably to a defect in the 'keratogenous zone' of the hair follicle, situated above the dermal papilla. Although the pattern of distribution of waved hairs at a gross level proved impossible to ascertain, individual zig-zag hairs could be scored both for the presence of an agouti band and for the waved phenotype (McLaren, 1971).

If both characters were determined by the same component of the hair follicle, a clonal theory would predict that the non-agouti hairs in the chimaeras should be waved, while the agouti hairs should be normal, as in the component strains. If the characters were determined by different clonal populations assorting randomly with respect to one another, e.g. mesoderm and ectoderm, the distributions of waviness and agoutiness should be unrelated. If one character were determined clonally, and the other systemically, their distributions should again be unrelated. Examination of individual hairs (McLaren, unpublished observations) showed no tendency for the proportion of waved hairs to differ as between agouti and non-agouti hairs, within a single patch of hairs. On the other hand if different patches within the same animal were sampled, making allowance for the different proportions of the two phenotypes in the animal as a whole, a significant *inverse* relationship ($P < 0.05$) between the proportion of waved and the proportion of non-agouti hairs was seen (Fig. 22*b*).

Why should this be? One possibility is that both characters are determined systemically, responding to some environmental variable. For example, low concentrations of sulphydryl groups are known to discourage the expression

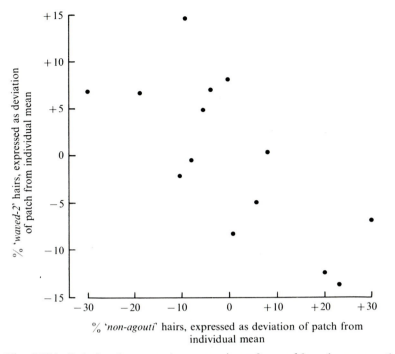

Fig. 22(*b*). Relation between the expression of *waved*-2 and *non-agouti* in hairs from follicles of chimaeric mice produced by aggregating homozygous *waved*-2 *non-agouti* embryos with embryos carrying the corresponding dominant allelles. Comparing patches in any one animal, the proportion of 'waved-2' hairs tends to be inversely related to the proportion of 'non-agouti' hairs.

of yellow pigment by melanocytes of any *agouti* locus genotype maintained in organ culture (Cleffman, 1963); the expression of the waved phenotype might possibly also be inhibited in the presence of low concentrations of sulphydryl groups.

Skin and other areas

Melanocytes are distributed freely in skin, as well as in hair follicles. Tarkowski (1964*b*) examined skin from several regions of the body of newborn agouti↔pink-eye dilution aggregants (ears, soles of feet, tail, area between genital papilla and anus), and found melanocytes from the agouti component in all cases where the pigmented retina was either fully pigmented or showed evidence of chimaerism. Unfortunately this result throws no light on the chimaeric status of the skin, since the presence of melanocytes of the other genotype could be neither proved nor rejected.

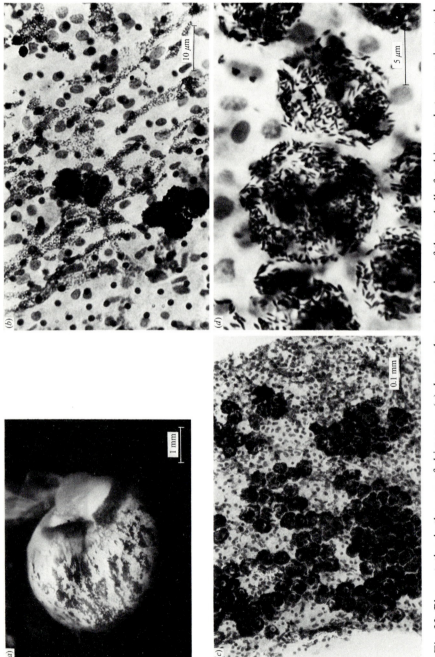

Fig. 23. Pigmentation in the eyes of chimaeras. (*a*) shows a low power view of the eyeball of a chimaera between a pigmented and a non-pigmented strain, with a higher power view (*b*) showing two small patches of pigmented cells in the retina, one containing two and the other four cells, overlain by streaks of extracellular melanin granules in the outer choroid layer. (*c*) and (*d*) show the pigmented retina, in which the granules remain within the cells; in (*c*) and (*d*) the choroid layer has been removed.

TABLE 9. *Numbers of cells per two-dimensional clone, in the pigmented retina of mice of various mosaic and chimaeric genotypes, estimated from one-dimensional sections of the retina*

	Left eyes		Right eyes	
Strain combination	No. of eyes	Mean no. of cells per clone ± standard error	No. of eyes	Mean no. of cells per clone ± standard error
X-inactivation mosaics ('flecked')	10	5.01 ± 0.45	10	4.23 ± 0.27
Unpigmented-Q ↔ pigmented-Q	6	5.81 ± 0.79	5	6.34 ± 0.29
Recessive ↔ pigmented-Q	8	5.91 ± 1.04	8	5.48 ± 1.18
Recessive ↔ (C57BL × C3H)F_1	10	8.37 ± 1.00	9	9.31 ± 1.24

The recessive ↔ (C57BL × C3H)F_1 chimaeras show significantly larger clones than any of the other groups. (From West, 1976*a*.)

Albino↔coloured chimaeras were reported by Deol & Whitten (1972*b*) to show melanocyte chimaerism in the iris, ciliary body and choroid layer of the eye. Eight very small regions of the inner ear were examined in the same animals: pigmented melanocytes were absent in more than 50 % of cases, suggesting that each region may have originated from a single cell (see Chapter 10), but the presence or absence of unpigmented melanocytes from the remainder could not be ascertained with certainty.

In the choroid, pigment granules are secreted extracellularly, so that although chimaerism may be grossly detectable, the distribution of the two components cannot be localized to the cellular level (Fig. 23). In the pigmented layer of the retina, on the other hand, the pigment granules remain within the cells, so that each cell can be identified unambiguously as pigmented or unpigmented.

Pigmented retina

The melanocytes of the pigmented epithelium of the retina differ from those elsewhere in the body in that they do not migrate from the neural crest, but are formed *in situ*, in the outer layer of the optic cup derived from the optic vesicle, which arises from the side wall of the fore brain. Pigment first appears within them on the 12th day of gestation. They form a single layer, so that a histological section through a chimaeric retina gives a one-dimensional string of pigmented and non-pigmented cells.

The first evidence that the two cell populations in aggregation chimaeras were finely interspersed came from studies of the pigmented retinas in new-born mice produced by aggregating pink-eye and normal embryos (Tarkowski, 1963, 1964*b*). Mintz & Sanyal (1970) and Mintz (1971*b*) mention that the retinas in albino↔pigmented chimaeras show pigmented and non-pigmented sectors, interpreted as 'radiating clones' of cells. Deol & Whitten (1972*a*) counted the numbers of pigmented and non-pigmented patches in sections of the retinas of albino↔pigmented chimaeras, for comparison with similar counts on the eyes of 'flecked' X-inactivation mosaics. The chimaeras showed big differences in the proportion of pigmented cells from one animal to another, and even in different parts of the same eye, though the left and right eyes were strongly correlated.

This type of analysis has been extended by West (1976*a*), who has measured the length of successive patches of pigmented and non-pigmented cells in sections of the retinas of chimaeric and mosaic mice killed at different ages. From such data the proportion of pigmented cells in each retina can be estimated, and hence the average number of clones expected to contribute to each patch, on the assumption of random distribution. (A discussion of the relation between clone size and patch size in chimaeras is given by West (1975). See also Chapter 10.) The number of cells per patch can also be calculated. For within-strain albino↔coloured chimaeras, the estimated average number of clones per patch at $12\frac{1}{2}$ days of embryonic development was almost exactly equal to the observed average number of cells per patch, suggesting that the immediate history of the tissue had been one of extensive cell mingling rather than coherent clonal proliferation. Adult animals of the same type showed five to six cells per clone in the retinal epithelium, and an increase of about threefold in the total number of clones per retina. This would be consistent with three to four cycles of cell division during development, together with a small amount of further cell mingling. In chimaeras of another strain combination a significantly larger patch size was found in the adult, which may indicate some tendency for differential cell adhesion (Table 9). This study, and that of Deol & Whitten (1972*a*), will be considered further in Chapters 9 and 10.

6

Other morphological characters

Other systems that have been examined in chimaeras include the skeletal system, the musculature and the neural retina.

Skeletal system

At a gross level, chimaeras between a normal strain and one carrying genes affecting skeletal and other morphological characters closely resemble the corresponding heterozygotes. Examples are seen in a series of chimaeras made between embryos homozygous for *short ear* (*se*/*se*) and *vestigial tail* (*vt*/*vt*), and the corresponding double dominants (McLaren & Bowman, 1969; McLaren, unpublished observations). In *se*/*se* homozygotes the external pinna is greatly reduced in size. Almost all the chimaeras had ears within the normal size-range; in this they resembled *se*/+ animals, since in its effect on the ears *short ear* is fully recessive. In *vt*/*vt* animals the tail is reduced to a mere stump, with variable numbers of caudal vertebrae heavily ankylosed with each other and with the sacrum. Of twenty chimaeras examined, sixteen had completely normal tails while four showed tail anomalies, ranging from very slight to moderately severe, but in no case approaching the homozygous condition. Heterozygous *vt*/+ mice also sometimes show slight tail anomalies. At another locus affecting tail development, chimaeras between T/t^x (tail-less) and +/+ (normal) embryos showed the short-tailed brachyury phenotype characteristic of $T/+$ heterozygotes (R. Pollard & D. Bennett, personal communication).

In homozygous form, *short ear* has a mild effect on many different parts of the skeleton, causing reduction of the number of ribs, bifurcated xiphisternum, absence of the omosternal elements, fusions between sternebrae, reduction of the ulnar sesamoids and other minor abnormalities (Green, 1968). When these characters were examined (Grüneberg & McLaren, 1972), the chimaeras proved to be more variable than *se*/+ heterozygotes and also to show considerably more mutant manifestation. The chimaeras were graded with respect to coat colour; the individuals manifesting *vt* and *se* more often and more strongly proved, with one exception, to be those in which the corresponding component was more strongly represented in the melanocyte population. A strong correlation was also found between the various skeletal

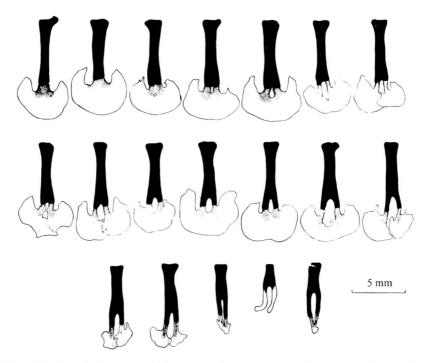

Fig. 24. The xiphisterna of nineteen chimaeras arranged in ascending order of abnormality. In the first row, one normal and six with a varying degree of involvement of the xiphoid cartilage (white) and its calcified granules (stippled); the second row has mild and the third more marked involvement of the osseous xiphisternum (black). From Grüneberg & McLaren (1972).

anomalies produced by *short ear* and *vestigial tail* in different parts of the body. If the genes act locally, this correlation suggests that the two chimaera components were distributed fairly evenly in the skeletal primordia.

With respect to bifurcation of the xiphisternum, 18/19 chimaeras showed abnormalities, compared with only 28/70 *se*/+ heterozygotes. The degree of abnormality was also greater in the chimaeras, encompassing the whole spectrum of manifestation except for the extreme homozygous condition (Fig. 24).

The ulnar sesamoid bone in the foot is either absent or very much reduced in *se/se* animals. The size of the bone shows a high degree of symmetry between right and left sides in +/+ and *se*/+ animals. In chimaeras (Fig. 25), the right–left symmetry is less marked, and varying degrees of intermediate expression are seen. The asymmetry suggests that bone size is determined locally, by the cell populations present in the bone rudiment, rather than by a systemic (e.g. hormonal) effect determined elsewhere in the body. The fact that some correlation between sides remains is probably due to variations in

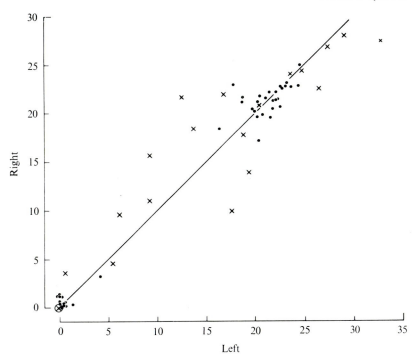

Fig. 25. The size (in cm², based on camera lucida drawings at magnification × 70) of the ulnar sesamoids in two strains of mice and in chimaeras between them. The two strains (indicated by dots) form two non-overlapping clusters, with the circle at the bottom left corner of the diagram representing twenty-four animals of the 'recessive' component in which the bone was absent symmetrically. The chimaeras (indicated by crosses) are less symmetrical than the component strains, and cover the whole size range. Some of the chimaeras were very old when killed and therefore had bones that were even larger than those of control animals of the 'dominant' component. From Grüneberg & McLaren (1972).

the initial proportions of the two cell populations making up the chimaeric embryo. The wide range of expression suggests that both populations are present to varying degree and that both exert an effect on the final phenotype.

The *short ear* gene acts in an essentially quantitative manner, reducing the rate of proliferation of the pre-cartilage cells and so producing a skeleton of reduced size (Green, 1968). Any qualitative differences that are observed are likely, therefore, to be due to threshold effects: when a cartilaginous blastema is reduced below a certain threshold level, ossification fails and the bone (e.g. the ulnar sesamoid) is completely absent. In heterozygotes, where both alleles are present in the same cell, some interaction between alleles must occur, to account for the partial dominance of the normal allele. In chimaeras, where the alleles are in different cells, the dominance relationship

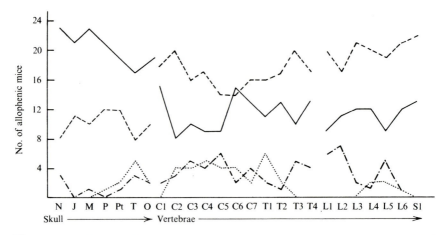

Fig. 26. Distribution of C57BL (———), C3H (– – –), asymmetric (. . . . .) and indeterminate (– · – · –) features in the skulls and vertebral columns of 35 C57BL↔C3H chimaeras (allophenic mice). From Moore & Mintz (1972).

is still maintained to a considerable degree, but many more animals show some mutant expression (e.g. 18/19 for the xiphisternum). This suggests that less allelic interaction occurs between cells than within cells: in chimaeras, whole patches of a tissue will consist of a single genotype, and in the absence of intercellular interaction this may be expected to give rise to an intermediate phenotype. If interaction involves some substance with limited powers of diffusion, a larger patch size will result in a phenotype more nearly intermediate between the mutant and the normal. This point will be taken up again in Chapter 9, when chimaeras are compared with females heterozygous for X-linked genes.

A consideration of the dental morphology of chimaeras involving the X-linked gene *tabby* will also be deferred to Chapter 9.

When threshold mechanisms and intercellular interactions are involved, as in the example of *short ear*, the skeletal phenotype does not directly reflect the underlying cellular patchwork formed as a result of cell lineage. Interstrain differences in vertebral column and skull development appear, however, to behave largely additively in chimaeras, with little indication of interaction, judging by the elegant clonal analysis carried out by Moore & Mintz (1972) on thirty-five C57BL↔C3H chimaeras.

The two component strains differed in the shape of the vertebrae. Each half-vertebra, on left or right side, was classified as C57BL, C3H or (rarely) indeterminate. Most but not all of the vertebrae were symmetrical on left and right sides, and there were also strong correlations between neighbouring vertebrae, though occasionally a single vertebra differed in type from both its anterior and posterior neighbour. A similar analysis was applied to the bones of the skull.

Although there was considerable variability from one chimaera to another in the relative proportions and distributions of the two components, certain regularities emerged. The C57BL type appeared more frequently in the skull and the C3H type in the vertebral column, especially in the lumbosacral region (Fig. 26). Moore & Mintz interpret this distribution as indicating 'an autonomous, strain-specific selective advantage enjoyed by C57BL/6 sclerotome cells in the early-forming (anterior) somites and by C3H in the later-forming (posterior) somites'. They point out that their observations agree well with a clonal model of vertebral column and skull development, in which each vertebra forms from a minimum of four separate clones or cell lineages. This follows directly from two facts: the asymmetry, showing that left and right sides are separately derived, and the indeterminacy, showing that more than one clone may be responsible for each half-vertebra. The model that they favour derives each vertebra intersegmentally from four clones, corresponding to the cranial and caudal sclerotome elements of neighbouring pairs of somites: right cranial, left cranial, right caudal, left caudal. They stress that four is a minimum estimate, since each element might contain several clones.

Musculature and neural system

We have already mentioned, in Chapter 2, the use of chimaeras to establish the origin of syncytial myotubes in skeletal muscle (Mintz & Baker, 1967). Since muscle taken from chimaeras between strains differing for electrophoretic variants of IDH showed the hybrid form of this dimeric enzyme, the syncytium must have arisen by fusion between myoblasts, some derived from one component and some from the other.

In early development, muscle is formed from the myotome element of the somites. Gearhart & Mintz (1971, 1972a) isolated individual somites from 8–9-day embryos chimaeric for different electrophoretic variants of the enzyme GPI. Thirty of the thirty-eight somites analysed were found to contain both GPI variants, showing that single somites were derived from more than one precursor cell. The proportions of the two components varied from one somite to another within the same embryo, but small groups of two to three neighbouring somites tended to have similar proportions. Not only the somite but also the smaller myotome element of the somite, was shown to be derived from more than one precursor cell, since individual eye muscles, each of which derives from the myotome of a single somite, also proved to contain both GPI variants when removed from chimaeras at 2–9 weeks of age (Gearhart & Mintz, 1972a). Of thirty-three extrinsic eye muscles analysed from chimaeras, all contained both components. Two-thirds also contained hybrid enzyme. Some implications of these findings will be considered further in Chapter 10.

Mintz (1972*b*) has summarized the information on somite origin that can be derived from the studies undertaken by her and her colleagues. The myotome of the somite, that gives rise to muscle, comes from at least two precursor cells or clones; the sclerotome, from which a half-vertebra is formed, may arise from a further two cells or clones; and the dermatome of the somite comes from one precursor cell responsible for a single hair follicle clone, and perhaps other precursor cells forming other dermal elements in the skin.

An interesting study of muscular dystrophy in chimaeric mice derived from aggregation of normal and genetically dystrophic embryos has been reported by Peterson (1974). A dystrophic allele was used that produced a form of the disease progressing slowly enough to allow mating, so that homozygous dystrophic embryos could be incorporated into the chimaeras. The constitution of individual chimaeric muscles could be ascertained by electrophoresis, since the chimaera components differed at the *Mod-1* locus and hence carried different isozymes of malic enzyme; the muscles were examined histologically; tests of muscle function were carried out; and the behaviour of the animals was recorded. No correlation between the genotype of a muscle and its dystrophic characteristics was found. In behaviour, the chimaeras were indistinguishable from normal animals of the same age, with none of the usual features of muscular dystrophy, and the twitch tensions in the muscles of the chimaeras proved to be at least as large as in the controls. Histological examination showed only very minor pathological features in the muscles; these were seen in genetically normal as well as in dystrophic muscle. On the other hand, the proportion of nuclei derived from the dystrophic component was substantial, averaging 58 % over forty-seven muscles from five chimaeras, and rising as high as 94 % in individual muscles. It seems clear that the primary genetic defect in murine muscular dystrophy is located not in the muscle itself, but elsewhere in the body, presumably in the nervous system. As with other characters situated at several removes from the site of primary gene action, the dominance relation in chimaeras is similar to that found in F_1 crosses.

When chimaeras were made between normal embryos and embryos homozygous for *reeler* (*rl*), a gene causing defects of balance due to pronounced degeneration of the cerebellum, three *rl/rl↔+/+* animals were entirely normal in their behaviour. A fourth, classified as *reeler* on behaviour, contained at least 95 % of the reeler component in its coat colour (R. J. Mullen & R. L. Sidman, personal communication). This again suggests that mutant characters behave as 'recessives' in chimaeras as in heterozygotes, and do not manifest unless the great majority of the cells are of the mutant genotype.

To examine a chimaera's behaviour, or muscle physiology, or even the

Fig. 27. Section of the retina of a C57BL↔C3H chimaera, showing an area (to the left) in which photoreceptor cells have degenerated so that only one or two rows of photoreceptor nuclei (PN) remain. The number of rows increases to the normal 8–10 on the right. The rod outer segments (OS) decrease in length along with the reduction in nuclei. From Wegmann, LaVail & Sidman (1971*b*).

structure of a bone that develops from a cartilaginous blastema, is to focus attention on a character several steps away from the underlying tissue patchwork. To throw light on cell lineage a gene is required with effects that can be identified in individual cells, and that acts strictly intracellularly. Few such genes are known for the mammalian nervous system. A possible candidate is the recessive gene for retinal degeneration (*rd*), homozygous in the C3H strain, that causes partial or complete blindness through postnatal degeneration of the neural retina. Chimaeras between C3H and C57BL or other genetically normal embryos showed areas of degeneration in the retina (Fig. 27), interspersed with areas of partially affected or normal tissue (Mintz & Sanyal, 1970; Mintz, 1971*c*; Wegmann *et al.*, 1971*b*). The existence of intermediate areas raises the suspicion that *rd* may not act strictly intracellularly, or alternatively that cell migration occurs to a significant degree (West, 1976*d*). Mintz & Sanyal report that the normal and degenerating patches conform to a basic plan involving ten radiating sectors per retina; they identify each sector as a single clone. When chimaeras were made involving differences in the pigmented retina as well as in the neural retina, no correspondence was seen in the quantitative proportions and location of the areas of albinism and of degeneration in a single eye, suggesting that, as would seem plausible on embryological grounds, the two layers involve separate cell lineages.

7

Immunology and blood

The very existence of a chimaera poses an immunological question. How can two populations of cells live together in apparent harmony when they are known to differ genetically and therefore presumably antigenically?

Antigenic chimaerism

That chimaeric individuals are indeed antigenically composite was first established by Mintz & Palm (1965), for the haematopoietic system. In a series of mice derived by aggregating embryos from two strains, C3H and C57BL, differing at the major histocompatibility locus H-2 (H-2^k and H-2^b respectively), agglutination tests with isoantisera revealed that some animals had both types of red blood cells, i.e. the cell population reacted with both anti-H-2^k and anti-H-2^b sera.

In a more extensive study of the same strain combination, Mintz & Palm (1969) examined several markers, including not only erythrocyte H-2 type, but also serum gamma-globulin allotype. Of twenty-six C3H↔C57BL mice proved to be chimaeric on one or other of the criteria examined, sixteen showed two populations of red blood cells by agglutination and sometimes also by absorption tests. Red blood cells and gamma-globulin showed some concordance, in that all but one of the mice with two erythrocyte populations also showed mixed C3H and C57BL gamma-globulin, while half the chimaeras with single red cell populations had only one type of gamma-globulin, corresponding genetically to their red cells. The techniques used were not sufficiently quantitative to relate the proportions of the two types of red cell and gamma-globulin in animals that had mixed populations of both.

The results suggested that C57BL erythropoietic tissue might possess some selective advantage over C3H, since only two animals were found with exclusively C3H red cells, and ten with exclusively C57BL. In the mixed populations also, C57BL red cells appeared to predominate, and one animal showed an increase with time in the relative number of C57BL red cells. The C57BL advantage seemed greater with respect to erythrocytes than gamma-globulin-producing cells.

If the two stem cell populations differ in their rate of proliferation not only during embryonic life, but also in the adult, a shift with time in the relative proportions of the two components in the blood is to be expected. The preponderance of C57BL over C3H red cells and its increase with time have recently been confirmed on an independent series of mouse chimaeras (West, 1976*b*), using a haemoglobin marker identified by starch gel electrophoresis. Tucker *et al.* (1974) found a striking change in the relative proportions of the component red cell populations in three sheep injection chimaeras, when the blood groups were examined at different ages. The blastomere donor type declined with age, even when it was initially in the majority. One sheep also showed a shift of transferrin type.

An extreme example of selection for erythrocytes of one genotype is provided by the two chimaeras described by Mintz (1970*b*), between W/W and $+/+$ embryos. The W/W genotype is normally lethal, so that the mouse dies of severe macrocytic anaemia a few days after birth. Although in the two $W/W \leftrightarrow +/+$ chimaeras the W/W component predominated in most tissues (e.g. liver, kidney, muscle), all the blood cells were $+/+$. No macrocytic anaemia developed, but both animals suffered from other mild blood disorders, due perhaps to some deleterious effect of the W/W component.

Wegmann & Gilman (1970) examined a further series of C3H↔C57BL aggregants, using the electrophoretic haemoglobin marker and an immunological test for 7 S gamma-globulin. The nine animals judged from their coat colour to be chimaeras all contained haemoglobin and gamma-globulin of both component types, but no great preponderance of C57BL red cells was seen, and no increase of C57BL cells with time.

The existence of antigenic chimaerism in solid tissues was established by grafting skin from C3H↔C57BL chimaeras onto recipients of the two component strains (Mintz & Silvers, 1967). Most of the grafts showed partial rejection, with necrotic and intact tissues intermingled. All but one of the thirteen coat colour chimaeras tested showed at least some graft survival on each component strain, demonstrating that chimaeric skin, like blood, contains two cell populations carrying different histocompatibility antigens.

Chimaerism in the lymphomyeloid system

The immune system in mammals includes bone marrow, spleen, thymus, Peyer's patches, lymph nodes, and circulating lymphocytes. We have already seen that chimaeras are capable of producing more than one type of gamma-globulin, corresponding to their component genomes. To what extent can chimaerism be detected elsewhere in the immune system, and can the proportions of the two components in the various tissues throw light on their developmental relationships?

The three relevant bodies of data published so far all confirm that the lymphomyeloid system can indeed be chimaeric, but the developmental

6

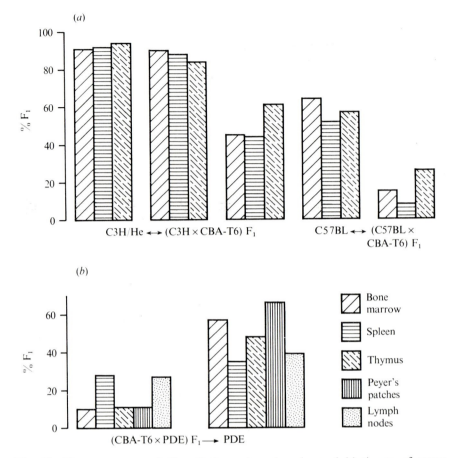

Fig. 28. The percentage of F_1 cells in various lymphomyeloid tissues of seven chimaeric mice. (*a*) data from Gornish, Webster & Wegmann (1972); (*b*) data from Ford, Evans & Gardner (1975*b*).

patterns that emerge are by no means uniform. Gornish, Webster & Wegmann (1972) examined mitotic cells in bone marrow, spleen and thymus from five aggregation chimaeras between either C57BL and (C57BL × CBA-T6)F_1 or C3H/He and (C3H × CBA-T6)F_1 embryos, and found a striking concordance between the proportions of cells carrying the T6 chromosome in the various tissues of a given animal. Their results, illustrated in Fig. 28(*a*), have been taken as support for the view that lymphomyeloid tissue throughout the body arises from a single stem-cell pool.

On the other hand, two injection chimaeras of the (CBA-T6 × PDE)$F_1 \rightarrow$ PDE combination, subjected to extensive chromosomal analysis by Ford *et al.* (1975*b*), gave results which suggested that bone marrow, thymus and

Peyer's patches were developmentally very closely related, while spleen and the five lymph nodes examined formed a different but almost equally homogeneous population. The data are given in Fig. 28(*b*). Circulating lymphocytes appeared to be different again, but this result may be complicated by the fact that the blood was cultured, and the two genetic populations of lymphocytes could react differently to in-vitro conditions or to stimulation with phytohaemagglutinin (PHA).

The apparent contradiction between these two sets of findings may be less than at first sight appears. Ford *et al.* emphasized the high variance, making the valid point that different tissues differ significantly from one another; Gornish *et al.*, by comparing between-mouse to within-mouse variance, were making the equally valid point that different chimaeras differ significantly. 'Founder cell effects' may contribute to heterogeneity between tissues: observations using cells marked by radiation-induced chromosome abnormalities have shown that a single stem cell colonizing the thymus can contribute a substantial proportion of the total cell population. Several studies on lymphomyeloid 'trafficking', and the fate of pre-programmed subpopulations, have been carried out since the original demonstration that bone marrow cells transferred to lethally irradiated recipients populated first the host bone marrow, and subsequently colonized the lymph nodes (Micklem, Ford, Evans & Gray, 1966), while transferred lymph node cells gave rise to a permanent lymph node population but never colonized the bone marrow (Ford, Micklem & Ogden, 1968).

The marked lack of conformity between spleen and bone marrow in the proportions of XX and XY cells in the sex chimaeras and mosaics studied by Mukherjee & Milet (1972) may be due to the fact that the spleen cells were cultured for an unspecified period before examination, and in several animals lacked one cell line entirely.

When two genetically distinct cell populations co-exist, cell selection due to differential rates of proliferation may also be of importance. It appears from Fig. 28(*a*) that a selective advantage of F_1 over C3H cells may be responsible for some of the uniformity between bone marrow, spleen and thymus found by Gornish *et al.* Selection is also strongly implicated in the studies carried out by Tuffrey, Barnes, Evans & Ford (1973*a*) and Ford *et al.* (1974) on the immune system of AKR↔CBA-T6 chimaeras. Although in the coat colour and in the germ cells the two component strains were more or less equally represented, in the lymphomyeloid system AKR cells appeared to be at a considerable advantage. In PHA-stimulated blood cultures the predominance of AKR mitoses was overwhelming at first sampling; when the mice were older a higher proportion of CBA cells was found, but the effect of variations in the culture technique could not be ruled out. In direct preparations CBA cells only made up 4 % of bone marrow, 10–13 % of spleen and thymus, and 20–21 % of lymph nodes and Peyer's patches. In this strain combination bone marrow, thymus and Peyer's patches, far from being homogeneous, were all different, as were spleen and lymph nodes. Presumably the selective advantage of the AKR genotype varied from tissue to tissue and determined the balance between the two cell populations, outweighing any initial resemblances due to stem-cell origins, and any differences due to founder effects. The AKR strain is known to be highly susceptible to spontaneous thymus-associated lymphoma (see Chapter 8).

A further complication is introduced by the observations of Bona *et al.* (1974) on the same series of AKR↔CBA chimaeras. When the two lymphocyte populations were distinguished not by their chromosomes but by their theta antigens, using immunoautoradiography, the proportions were much more equal, ranging from 24 % to 76 % AKR, with a mean of 46 %. In contrast to control F_1 animals, no cells in the chimaeras carried both antigens, implying that cell hybridization seldom if ever occurs. The contradiction between the theta antigen and the chromosomal results could be reconciled if the high proliferative rate of AKR cells in response to PHA *in vitro* were associated with a relatively low rate of spontaneous proliferation *in vivo*, or if a high rate of spontaneous proliferation were balanced by a high rate of cell death. In either case, the antigenic analysis would provide a more accurate picture of the actual balance of cellular forces in the tissue than the chromosomal studies. The possibility remains, however, that the theta phenotype of a cell does not always reflect its genotype.

The conclusion that little or no cell hybridization occurs within the lymphomyeloid complex is further supported by observations on two rabbit chimaeras (Gardner & Munro, 1974). When allotype markers on the immunoglobulin heavy and light chains were examined, no mixed molecules were found, with a limit of detection of 1/300 (A. J. Munro & R. L. Gardner, personal communication). A similar study on allotypes in chimaeric mice from strains differing at the *H-2* locus suggests that somatic cell hybridization does sometimes occur (Munro, Day & Gardner, 1974), but the result requires confirmation and other explanations of the data are possible.

Tolerance

We have seen that the two component cell populations of a chimaera retain their antigenic specificity, and that the chimaerism extends to the immune system. Since the animals are healthy and long-lived, it follows that each component must be in some way immunologically 'tolerant' of the other. The mechanism of this tolerance is still not known.

In chimaeras made from inbred embryos the tolerance is specific, in that it extends to tissues from other individuals of the component strains, but not to unrelated strains. Thus chimaeras of several strain combinations were shown to accept skin grafts from the two component strains, but rejected grafts from 'third-party' strains (Mintz & Silvers, 1967). Occasional animals rejected skin from one of the component strains, although the presence of cells from that component could be demonstrated, e.g. in bone marrow (Mintz & Palm, 1969), suggesting that some antigens present in skin may be absent from bone marrow. No evidence of immunological runting was seen, and the successful grafts remained intact throughout the host's life-time, sometimes till $2-2\frac{1}{2}$ years of age. The authors designated the tolerance 'intrinsic' rather than 'acquired', and proposed that the mechanism by which it arose was identical to that involved in the establishment of 'self-recognition' and 'self-tolerance' during normal development, the 'self' in the case of a chimaera being composite. 'Association of genetically unlike cells prior to differentiation precludes any subsequent appraisal by either type of the other as foreign' (Mintz & Silvers, 1970).

This view is in accordance with the 'forbidden clone' concept of Burnet (1959), which postulates that when a clone of cells arises that is capable of recognizing and reacting against the animal's own tissues (i.e. 'self'), that clone is eliminated.

However, when lymph node cells from C57BL↔C3H/He or SJL↔ (CBA × C57BL)F$_1$ aggregation chimaeras were examined by a microcytotoxicity test, a reaction against both component cell types was claimed (Wegmann et al., 1971a). The destruction in vitro of component strain fibroblasts by cells from chimaeras was more effective than when lymph node cells from either component strain or their F$_1$ hybrid were used, and resembled that mediated by cells from immunized donors. The reaction could be prevented by serum from the chimaeras, but not from F$_1$ or component strain mice, suggesting that chimaeric 'tolerance' involved mutual sensitization of the two immune systems, along with some blocking effect of the serum, rather than any central failure of the immune response. The concept is illustrated in Fig. 29.

Similar results have been obtained in another assay system, mixed lymphocyte culture, using cells from C57BL↔SJL chimaeras (Phillips, Martin, Shaw & Wegmann, 1971). In contrast, spleen cells from C57BL↔C3H/He chimaeras did not respond in vitro to cells from the component strains, and indeed were able to prevent component-strain spleen cells from reacting in mixed lymphocyte culture either to one another or to the chimaeric cells (Phillips & Wegmann, 1973). Thus blocking appears to be an active phenomenon. Since it was immunologically specific, and was unaffected by anti-theta antibody, it could have been mediated by B lymphocytes, or by plasma cells. The degree of suppression exerted on cells from each component strain varied widely from one chimaera to another.

The apparent difference between chimaeras, where the genetically contrasting histocompatibility alleles are located in different cells, and F$_1$ animals, where they co-exist in the same cell, suggests that a lymphocyte may be

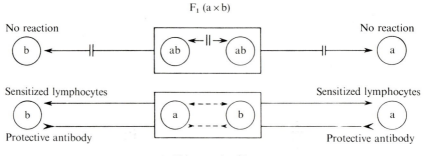

Fig. 29. A comparison between the immunological situation in F_1 mice from a cross between two strains a and b, and in chimaeras between a and b. The central boxes portray the situation *in vivo*, and the outer areas the results of in-vitro tests. Normal arrow heads indicate antagonistic (e.g. cytotoxic) reactions, reverse arrow heads protective reactions, and vertical lines absence of a reaction.

unable to synthesize specific antibody against an antigenic determinant on its own membrane (Allison, 1971). Thus F_1 cells, that carry surface histocompatibility antigens of both parental types, would fail to respond to either.

This view envisages unresponsiveness at the cellular level as an autonomous phenomenon, intrinsic to the cell itself. The 'forbidden clone' theory, on the other hand, regards intercellular interactions as the basis for cellular unresponsiveness. Deol (1973) has pointed out that the findings on chimaeras are not necessarily inconsistent with the 'forbidden clone' theory, since component cell populations in chimaeras tend to be clumped, owing to clonal proliferation, rather than randomly intermingled (see Chapter 10). Within for example the thymus, patches of a single genotype may be sufficiently large and stable to allow many of the lymphoid cells to undergo immunological maturation in effective isolation from cells of the contrasting genotype, so that these are subsequently treated as foreign. Observations of sections of mesenteric lymph nodes from chimaeric mice stained with fluorescent anti-*H-2* antisera directed against the component antigens have, however, indicated that patch size in lymph nodes is small, with the two components closely intermingled (C. Kaushgen, M. Edidin and T. G. Wegmann, personal communication).

The idea of an immunological equilibrium between sensitization of lymphoid cells and protection by a blocking factor is not confined to chimaeras, but has also been invoked to explain the non-rejection by the mother of an immunologically incompatible foetus (Hellström & Hellström, 1970), and the development of tolerance in response to the injection of allogeneic cells into neonatal (Hellström, Hellström & Allison, 1971) or lethally irradiated animals (Hellström, Hellström, Storb & Thomas, 1970). The blocking factor is thought to be either an antibody or, more likely, an antigen–antibody complex, similar perhaps to the 'enhancing' factor demonstrated in tumour immunity (Kaliss, 1969). The antibody component might be directed specifically against the relevant receptor sites on the surface of immunologically competent cells (see Ramseier, 1973). An attempt to produce tumour enhancement using sera from chimaeras of an appropriate strain combination proved unsuccessful (Kaliss, Whitten, Wegmann & Carter, 1974).

Another situation in which unreactiveness to 'self' antigens breaks down, which may be analogous to the findings on chimaeras described above, is the spontaneous autoimmune disease of NZB mice. Russell, Liburd & Diener (1974) have reported that theta-positive lymphocytes from old NZB animals show in-vitro cytotoxicity to syngeneic fibroblasts. The reaction can be suppressed by lymphocytes from young mice: as in Phillips & Wegmann's report of active suppression by lymphocytes of chimaeras, the cells responsible for the suppression do not carry theta antigen and so are presumably B cells.

Recently, however, the interpretation of the immune status of chimaeras has been still further complicated by the failure of Meo, Matsunaga & Rijnbeek (1973) to confirm the earlier findings. They examined eight C57BL↔ C3H/He chimaeras, one of the strain combinations used by Wegmann and his colleagues, and found the lymphocytes totally unresponsive to component strain cells in mixed lymphocyte cultures, while no blocking action of the chimaeric sera on component strain cells could be detected. Similar specific unresponsiveness was reported in the lymphocytes of a human chimaera (Ceppellini, 1971). The two populations of lymphocytes showed no reactivity in mixed lymphocyte culture towards one another or towards other cells of equivalent *HL-A* type, and the chimaera's serum showed no blocking effect on the reactivity of its lymphocytes towards 'third-party' cells.

The results of Meo *et al.* may not be irreconcilable with those of Wegmann and his colleagues. The failure of spleen cells from chimaeras to respond *in vitro* to cells of either component was found not only by Meo *et al.* but also by Phillips & Wegmann (1973), who attributed it to active suppression exerted by chimaeric cells. A blocking effect of chimaeric sera in mixed lymphocyte cultures was found by Phillips *et al.* (1971) but not by Meo *et al.*; the strain combinations in this instance were not the same, and the SJL strain used in the earlier work is known to show various immunological idiosyncrasies.

The results with cell-mediated suppression of mixed lymphocyte cultures in the C57BL↔C3H strain combination were very variable (Phillips & Wegmann, 1973), suggesting that certain critical factors have not as yet been identified. Cells from some chimaeras suppressed the reaction of both component strains to varying degrees, some suppressed one only, and some neither. Perhaps the immunosuppressive ability of chimaeric cells is only demonstrable *in vitro* if it has been increased above the normal level, for example by immunological challenge. A possible model is provided by the demonstration that some form of active suppression involved in neonatally induced tolerance can be stimulated by challenge (Ramseier, 1973). CBA mice injected at birth with $(A \times CBA)F_1$ spleen cells were rendered tolerant to A cells; this tolerance could be broken if the adult animal was given large numbers of CBA cells sensitized against A, but not if a small dose of normal CBA cells had been administered earlier. Ramseier suggested that the earlier inoculum acted by boosting the levels of serum antibody with anti-receptor activity (i.e. antibody against CBA receptors for A antigen): no such serum factor could be detected in 'steady state' tolerant mice, but high titres were formed after challenge. Thus it could be that in 'steady state' chimaeras no blocking activity is detectable in the serum because all blocking antibody is engaged in blocking and no surplus is formed.

The strange case of the NZB chimaeras

Mice of the New Zealand Black (NZB) strain invariably develop a progressive autoimmune disease, characterized by Coombs-positive haemolytic anaemia and believed to involve a breakdown with age of some T-cell-mediated autoimmune control. To determine whether the presence of a normal auto-immune control mechanism would inhibit the development of chronic NZB disease, aggregation chimaeras were made between inbred NZB and non-inbred CFW embryos (Barnes *et al.*, 1972*a*). The authors claimed, on the basis of serological, haematological and morphological evidence (e.g. the absence of 'germinal centres' in the thymus), that NZB disease was indeed suppressed in the chimaeras. Three of the four chimaeras developed positive direct Coombs test reactions and other indications of acute allogeneic disease, but these were interpreted as indicating some different immunological reaction between the chimaera components.

The occurrence of immunological reaction ('graft-versus-host disease') in the NZB↔CFW chimaeras was first suspected when red blood cells from the chimaeras gave a positive reaction with NZB antiserum against CFW immunoglobulin allotype (direct Coombs test) but, unlike Coombs-positive NZB mice, failed to react with CFW antiserum against NZB immunoglobulin allotype (Tuffrey & Barnes, 1973). Since haemagglutination tests with specific H-2 antisera had established that NZB as well as CFW red cell populations were originally present in the chimaeras, the CFW allotype specificity of the direct Coombs test reaction suggested that the NZB red cells were coated with CFW immunoglobulin. The presumed CFW 'autoantibody', when eluted from the red cells of the chimaeras, reacted not only with red cells from old NZB mice, but also with cells from young NZB mice incubated with old NZB serum, and hence was probably directed not against the old NZB red

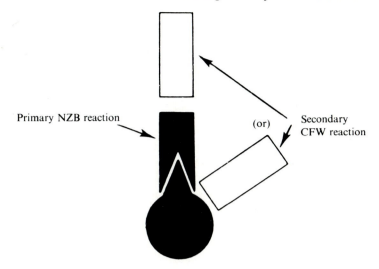

Fig. 30. Proposed sequence of events in NZB↔CFW chimaeras. An NZB auto-antibody (black) is formed against NZB red blood cells; CFW autoantibody (white) then reacts with the NZB antibody, or antigen antibody complex. From Barnes & Tuffrey (1973).

cell itself, but against an NZB red cell/NZB autoantibody complex (Fig. 30) (Barnes & Tuffrey, 1973).

A breakdown of tolerance in the NZB↔CFW chimaeras (Barnes & Tuffrey, 1972*b*) is supported by other changes that occurred as the animals aged. In one, the CFW red cell population disappeared, while another lost its NZB red cells and some of the black hairs in the coat, and acquired the ability to reject NZB skin (Barnes & Tuffrey, 1972*a*). These shifts in cell populations need not, however, have had an immunological basis. Serum haemolytic complement activity was absent (Barnes *et al.*, 1974*c*) as in the NZB strain, although CFW mice have normal levels of complement and chimaeras of strain combinations not involving NZB showed little or no reduction in complement activity. Deposits of complement were later detected in renal lesions, together with antibody. The pattern of immunofluorescent staining in the lesions closely resembled that seen in acute graft-versus-host disease. Excess removal of complement in complexes is known to accompany graft-versus-host disease, and could explain the absence of serum complement activity. A graft-versus-host assay based on popliteal lymph node weight gain proved positive.

Immunofluorescence techniques (Barnes *et al.*, 1974*a*) not only established the presence of both cell populations in thymus, spleen, lymph nodes and kidney of the chimaeras, but also revealed the phagocytosis of erythrocytes by monocytes and macrophages in all four animals. Immunoglobulin, both allotypes, complement and *H-2* antigens were also demonstrated in the macrophages.

The survival to 80 weeks of age of one NZB↔CFW chimaera, negative in the direct Coombs test and with no evidence of NZB or graft-versus-host disease in spite of the presence of NZB cells, is attributed by Barnes *et al.* (1974*a*) to active macrophage-mediated autoimmune clonal suppression, as electron microscopy showed marked phagocytosis of plasma cells. The Gross-antigen-associated type-C leukaemia virus, thought to be the primary stimulus to autoantibody formation in the NZB strain, was detected in all the chimaeras, though in least amount in the Coombs-negative animal in which autoimmune activity was most successfully suppressed (Barnes, Wills & Tuffrey, 1975).

The authors propose that the primary defect in NZB mice is a failure of T cell suppressor function, which in the chimaeras is corrected through elimination of the aberrant autoimmune clones by macrophage activity. The potential importance of such a mechanism for the control of autoimmunity in general is obvious.

Although NZB↔CFW chimaeras fail to develop NZB disease, they are fully capable of transmitting to their progeny the capacity to develop the disease (Tuffrey, Kingman & Barnes, 1973*b*).

Chimaeras as immunological test systems

Immunological techniques are of value in investigating chimaeras; the existence of chimaerism is in itself of immunological interest; in addition, chimaeras are of value in investigating immunological problems.

An example is the use by Mintz (1970*d*) of chimaeras as test animals to look for F_1 hybrid antigens. Skin from a (C3H × C57BL)F_1 donor was readily accepted by a C3H↔C57BL chimaera, suggesting that no hybrid-specific histocompatibility antigens were present in the F_1 skin.

A more complex problem concerns the reduced ability of certain strains of mice to form antibodies against the synthetic polypeptide (T,G)-A--L. The immune response gene involved is *Ir-1A*, which is closely linked to the *H-2* locus (McDevitt *et al.*, 1971). The gene was thought to act through T lymphocytes; it was not known whether it also affected B cells. Chimaeras between high-responding and low-responding strains (C57BL and C3H) included some high responders (Bechtol *et al.*, 1971). In a different strain combination (CWB↔C3H) most of the chimaeras proved to be high responders, and since the component strains differed at the *Ig-1* immunoglobulin allotype locus, it was possible to correlate antibody response with immunoglobulin type (Freed *et al.*, 1973). Most of the high responders were predominantly Ig-1b, like the CWB strain, but some animals had significant amounts of the C3H Ig-1a, and yet showed a high response. Examination of the allotype distribution of the specific antibodies showed that both Ig-1a and Ig-1b anti-(T,G)-A--L was produced (Bechtol, Freed, Herzenberg & McDevitt, 1974).

In a further study on C57BL↔C3H chimaeras (Bechtol *et al.*, 1974), the amount of low-responder-allotype anti-(T,G)-A--L in chimaeras was found to be four to five times greater than in control C3H mice. The proportion of *a* allotype in the specific antibody tended to be lower than in the total serum, suggesting perhaps that C57BL T cells and C3H B cells co-operate less well than do T and B cells of the same genotype (McDevitt, 1972). The correlation between the proportion of *a* in the specific response and in the total serum was much closer when congenic C3H mice were used as chimaera components (Bechtol *et al.*, 1974).

This result strongly supports the hypothesis that the *Ir-1A* gene is not expressed in B cells. An alternative interpretation, that the increased antibody response of the 'low-responder' B cells is due to non-specific stimulation resulting from immunological interaction between the chimaera components, has been ruled out by two further series of chimaeras (K. B. Bechtol, personal communication). By aggregating homozygous $H\text{-}2^{k/k}$ low-responder embryos (C3H) with heterozygous $H\text{-}2^{k/b}$ high-responder embryos (CKB × CWB), chimaeras were produced in which the high-responder cells shared the whole of the $H\text{-}2^k$ complex with the low-responder cells. High levels of low-responder-allotype anti-(T,G)-A--L were still found. In the second series, embryos of two histoincompatible low-responder strains were aggregated, C3H.Q ($H\text{-}2^{q/q}$) and CKB ($H\text{-}2^{k/k}$). Only low responses to (T,G)-A--L were produced, suggesting that a histoincompatibility reaction (allogeneic effect) in itself could not increase the response. Evidently the high C3H response in the earlier chimaeras was not due to abnormal stimulation of the low-responder-genotype B cells.

It has been suggested that the failure of 'low-responder' strains to form antibodies against (T,G)-A--L is due to the presence of an antigenically similar self-component in these strains. The finding that chimaeras between high- and low-responder strains are responsive might then be thought to support the view that the low level of autoimmunity in chimaeras is due to some form of 'balanced enhancement' (for discussion, see Cohn, 1972).

8

Tumours

Differences between strains in susceptibility to tumours, and in the incidence of particular types of tumours, have long been known. Little is known, however, of the physiological basis of these genetic differences, though in some instances an oncogenic virus has been found to be involved.

Experimental chimaerism offers a means of combining cell lines of differing tumour susceptibility within a single organism, so as to study the relations between factors associated with susceptibility, and those associated with resistance. Investigations so far have sought to determine the incidence of tumours in chimaeras between high- and low-tumour strains, the tissue specificity of the tumours that develop, and the extent to which they remain confined to the susceptible cell population. Once again we may ask whether the cell phenotype, in this case a malignant transformation, reflects the genotype of the cell, or its environment.

Mammary tumours

Female mice of the C3H inbred strain receiving mammary tumour virus in their mother's milk show a high incidence of mammary tumours (nearly 100 % at an average age of 10 months). The C3Hf strain has a low-virulence tumour virus transmitted in the germ line, and shows a lower incidence of tumours and a later age of onset. C57BL mice rarely develop spontaneous mammary tumours, though the incidence can be increased to about 40 % by fostering onto C3H females.

In a preliminary study of two C57BL↔C3Hf chimaeras (Mintz & Slemmer, 1969), grafting of tissue from single mammary glands onto either component strain was followed by partial rejection, suggesting that the epithelium of each chimaeric gland contained a mixed cell population. Pre-malignant hyperplastic nodules taken from the chimaeric glands also proved to contain cells of both components; histochemical staining for β-glucuronidase revealed both C3H and C57BL cells within a single alveolus (Mintz & Condamine, cited by Mintz, 1972a). Since one animal was an XX/XY chimaera, it seems that XY cells must be capable of responding to the female hormonal environment in such a way as to contribute to normal mammary gland epithelium. Graft outgrowths from the hyperplastic nodules gave evidence of strain-specific growth patterns, with C3Hf cells in C3Hf recipients continuing hyperplastic, and C57BL cells in C57BL recipients growing normally. The authors conclude that although C57BL cells are incorporated into characteristic pre-malignant nodules, they retain their inherent growth characteristics and are not affected by the partially C3Hf environment of the chimaera. However, the possibility also exists that the growth characteristics revert under the environmental influence of the recipient, after grafting; hence an earlier effect of the chimaeric environment cannot be excluded. This interpretation would seem to be supported by the observation (Slemmer & Mintz, cited by Mintz, 1972*a*) that chimaeric nodules grafted to F_1 recipients gave rise to tumours that also contained both components.

However, C57BL cells in C57BL↔C3Hf chimaeras rarely give rise to mammary tumours (Mintz, 1970*e*). Transferred to foster-mothers lacking mammary tumour virus, 5/10 C57BL↔C3Hf chimaeras developed tumours, at an average age of 18 months, compared with 6/14 at an average age of 15 months in C3Hf controls, and 0/9 in C57BL controls. The incidence of tumours was thus no lower in the chimaeras than in the susceptible component strain. Six tumours were analysed by grafting or enzyme determinations; five proved to be C3Hf, and one was C57BL. Reared in C3H foster-mothers carrying mammary tumour virus, 23/27 C57BL↔C3H chimaeras developed tumours, at an average age of 10 months, compared with 16/16 at the same average age in C3H controls, and 6/15, at an average age of 14 months, in similarly reared C57BL controls. Of forty-one tumours analysed, thirty-three were entirely C3H, five contained some C57BL, and three were entirely C57BL.

Thus the chimaeras resembled the more susceptible strain (C3H) in extent and rate of mammary tumour formation, and the tumours that developed were derived largely, but not exclusively, from the more susceptible cell population. In four of the five mixed tumours, the C3H component was very small, and could have resulted from vascular or connective tissue contamination; in the remaining tumour the two components were present in roughly equal amounts, suggesting that it may have arisen from more than one cell. In general, however, the tumours were of single type. The fact that most were C3H implies a large autonomous component in the expression of tumour phenotype; on the other hand the fact that tumours also developed from the C57BL component shows that autonomy was not absolute, and that other cells in the same gland or elsewhere in the body were capable of influencing tumour expression.

No correlation was seen between the development of mammary tumours and the amount of high-tumour component present in other tissues of the same individual (immunoglobulin, spleen, erythrocytes, lymph nodes), suggesting that these tissues were not involved to any significant extent in the genetic differences in tumour susceptibility. Mintz (1970*e*) points out that

C3Hf predominated over C57BL tissue in normal mammary glands of both female and male chimaeras, and suggests that the susceptibility to malignant transformation of C3Hf mammary cells may be related to their higher growth rate. A similar suggestion has been made for the lymphoma-susceptible AKR strain, the lymphomyeloid cells of which predominate over those of C3H and CBA origin (see below).

Hepatomas

The development of liver tumours was studied in males of the same series of C3H↔C57BL and C3Hf↔C57BL chimaeras that had yielded data on mammary tumours (Mintz, 1970*e*). C3H and C3Hf strains show a high incidence of hepatomas, while C57BL mice rarely develop these tumours.

In neither group was the frequency of spontaneous hepatomas among the chimaeric males significantly lower than in the more susceptible strain: 13/29 and 3/10 in C3H↔C57BL and C3Hf↔C57BL chimaeras respectively, compared with 9/13 for C3H, 3/9 for C3Hf and 1/19 for C57BL controls (Mintz, 1970*e*). The chimaeras also resembled the high-tumour strain in mean number of tumorous foci per liver. Enzyme determinations (using strain-specific electrophoretic variants of NADP-malate dehydrogenase) revealed twenty-four tumours of C3H or C3Hf type, one of C57BL, and two containing both C3H and C57BL cells. As with mammary tumours, it seems clear that there was a large element of autonomy in the expression of high and low tumour susceptibility in the chimaeras, with an overall tumour incidence little if at all below that of the susceptible strain. Most of the tumours derived from high-tumour cells, but there was at least one derived entirely from the 'resistant' strain, C57BL. Whether or not both components were tumorous in the two mixed tumours could not be ascertained, but in a later study two more mixed tumours were examined and malignant cells of both genotypes were identified histochemically (Condamine *et al.*, 1971).

Analysis of forty-one normal livers from male and female chimaeras (Mintz, 1970*e*) showed that, as in the mammary gland, C3H and C3Hf predominated over C57BL tissue, perhaps reflecting the tumorigenic potential of the strains. Forty-four per cent of the livers contained no C57BL tissue. When both components were present the genotype of a tumour showed little relation to the genetic composition of normal tissue in the same lobe. Histochemical visualization of the genotypic composition of both normal liver and hepatomas, using a β-glucuronidase marker such that C57BL or BALB/c cells stained red (positive) and C3H cells green (negative), confirmed that pure-strain tumours often arose from liver lobes that included cells of both genotypes (Condamine *et al.*, 1971).

Lung tumours

A similar picture is seen for lung tumours (Mintz, Custer & Donnelly, 1971). BALB/c mice have a relatively high incidence of spontaneous lung tumours (11/19); C57BL and C3H mice have a very low incidence (0/26). Chimaeras between high- and low-tumour strains showed a susceptibility (8/22) again less than that of the BALB/c strain, but not significantly so. Enzyme determinations revealed that the tumours arising in the chimaeras were largely BALB/c in composition, although nearby normal lung tissue often showed substantial amounts of the low-tumour component. More than half the tumours also contained cells from the low-tumour component, but whether or not these cells were malignant was not ascertained. The distribution of the two chimaera components in other tissues of the body bore no obvious relation to the incidence of lung tumours.

Leukaemia

The high incidence of thymus-associated lymphocytic leukaemia in AKR mice is related to the presence of oncogenic Gross virus, transmitted though the germ line. C3H and CBA strains are not normally susceptible to leukaemia; C3H mice that are infected with virus at less than 12 hours of age develop the disease in later life, but CBA mice remain resistant to leukaemia, even when derived from AKR foster-mothers by embryo transfer.

In a preliminary experiment, Mintz *et al.* (1971) found that three AKR↔C3H chimaeras all developed leukaemia. Solid lymphoid tumours, white blood cells and red blood cells proved to be uniformly AKR in constitution. Two of the animals also developed hepatomas: these were composed of cells of the high-hepatoma C3H component, although the liver parenchyma was extensively infiltrated with leukaemic AKR cells.

With a different strain combination, AKR↔CBA, the development of leukaemia appeared to be delayed or even suppressed in a proportion of chimaeras (Barnes, Tuffrey & Kingman, 1972; Tuffrey & Barnes, 1972). Control AKR mice all developed lymphomas by 56 weeks of age; two chimaeras developed lymphomas at 32 and 56 weeks of age, but three were still well at 68 weeks. A larger series of chimaeras confirmed these preliminary findings (Barnes, Tuffrey & Ford, 1973). Tumours were seen in fewer than half the animals, and the mean age of onset was significantly delayed (Fig. 31). Five of the six tumours analysed were AKR in origin, but one proved to be CBA, in spite of the fact that CBA cells in their normal environment are very highly tumour-resistant. This suggests some influence of the susceptible AKR cells on the CBA population.

Transfer of the chimaeric embryos to uterine foster-mothers was not the cause of the delay in tumour onset, as Barnes & Tuffrey (1974) found no

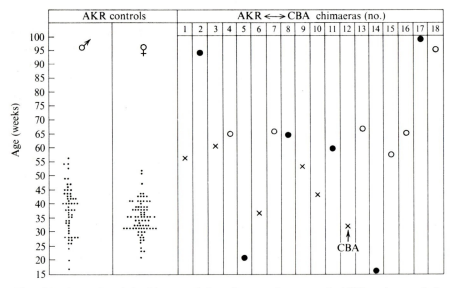

Fig. 31. Age-related incidence of lymphomas in normal AKR mice and in AKR↔CBA chimaeras. ○, alive and well; ×, dead with lymphoma; ●, dead with no lymphoma. The arrow indicates the lymphoma arising from the CBA component. After Barnes, Tuffrey & Ford (1973).

effect of transferring pure AKR embryos to CBA foster-mothers. Nor was there any shortage of AKR cells in the blood: on the contrary, cytogenetic analysis (using the T6 marker) of peripheral blood samples showed an over-whelming predominance (> 95 % on average) of AKR lymphoid cells (Tuffrey *et al.*, 1973*a*). This predominance of AKR cells was also seen (see Chapter 10) in lymphomyeloid and other somatic tissues (Ford *et al.*, 1974), but not in coat pigmentation nor in germ cells.

An alternative possibility, that the suppression of leukaemia in the chimaeras was due to absence or delayed development of the oncogenic virus, was ruled out by the identification of large numbers of typical type-C murine leukaemic virus particles in the chimaeras, derived presumably from the AKR germ cells (Wills, Tuffrey & Barnes, 1975). Their antigenic specificity as Gross virus was confirmed by immunofluorescence absorption assays of the chimaeric tissues (R. D. Barnes, personal communication).

Pre-leukaemic AKR mice are characterized by extensive antibody–antigen complexes in the renal glomeruli, formed in part by reaction of Gross viral antigen with specific anti-Gross-virus antibody. These complexes were markedly reduced in the AKR↔CBA chimaeras (R. D. Barnes, personal communication), suggesting that the chimaeras may owe their relative tumour resistance to the ability of even a small number of CBA cells to maintain tolerance to Gross virus. The susceptibility of AKR mice is thought to be due to a failure with age in the mechanism of tolerance to the virus: anti-viral antibodies are formed, and the resultant antigen–antibody complexes (detectable in the renal glomeruli) may serve to 'mask' tumour-specific sites, preventing detection and elimination of lymphomas. A search for specific tumour immunity in the chimaeras was unsuccessful (MacLennan & Hollingsworth, cited by Barnes, 1974): neither humoral nor cellular activity against AKR lymphoma cells *in vitro* could be detected, though the absence of in-vitro cytotoxicity does not necessarily exclude the possibility that tumours develop and are immunologically eliminated *in vivo*.

Conclusions

The studies on chimaeras have left little doubt that the type of tumour that develops usually depends on the genotype of the cell or cells undergoing malignant transformation. For example, in AKR↔C3H chimaeras, lymphomas are derived from the AKR and hepatomas from the C3H component. A few exceptions almost always occur, where the tumour proves to be derived from the 'resistant' component; perhaps the most surprising example is the CBA leukaemia found in an AKR↔CBA animal. A few tumours of apparently mixed origin have also been reported.

The extent to which overall tumour susceptibility may be affected by the cellular environment, or by systemic factors, seems to vary from one strain combination to another. Where both components are susceptible, though to different tumours, cells appear to function autonomously, and incidence of each type of tumour resembles that found in the corresponding component strain. On the other hand, a highly tumour-resistant strain, such as CBA, is apparently capable of influencing the incidence of detectable tumours in a more susceptible partner. From the point of view of analysing normal mechanisms of tumour immunity, the latter situation would appear the more potentially rewarding.

9

Chimaeras versus mosaics

As defined in Chapter 1, mosaics and chimaeras both contain two or more genetically distinct cell populations; they differ in that mosaics are derived from a single zygote, chimaeras from more than one zygote.

The mosaicism most intensively studied in mammals is the consequence of the inactivation of one X chromosome in each cell. Monotremes apparently do not show X chromosome inactivation; but in the somatic tissues of female marsupials and eutherian mammals, XX in chromosome constitution, each cell contains only a single genetically active X chromosome. Which of the pair of X chromosomes is to be active and which inactive is decided early in development. In marsupials the X chromosome derived from the father seems to be preferentially inactivated, but in eutherian mammals (with the possible exception of extra-embryonic tissues, see Takagi & Sasaki, 1975), inactivation is at random.

Once the decision has been taken, every descendant of the cell has the same X chromosome active. Thus a female eutherian mammal is automatically mosaic for any X-linked locus at which she happens to be heterozygous. To take the *tabby* (*Ta*) locus as an example, a *Ta*/+ female mouse will have the *tabby* gene active in approximately half her cells, and the normal wild-type gene active in the other half.

X-inactivation mosaicism is not confined to genes normally carried on the X chromosome, as X–autosome translocations exist in which a piece of autosomal chromosome is attached to an X, and activated or inactivated along with it. The most widely used is Cattanach's translocation, which consists of a segment of chromosome 7, including the wild-type alleles at the *albino* and *pink-eye* loci, translocated onto the X chromosome. A female mouse carrying two autosomal copies of say the *albino* gene and heterozygous for Cattanach's translocation will be a mosaic for albino and non-albino, since melanocytes with the translocation X active will express the dominant non-albino phenotype, synthesizing melanin, while melanocytes with the normal X chromosome active will express albino.

By aggregating embryos destined to become albino and non-albino, or tabby and normal, chimaeras can be constructed that are conceptually very similar to the corresponding X-inactivation mosaics. The analogy is drawn in Fig. 32. Similar though the two situations are, important differences exist

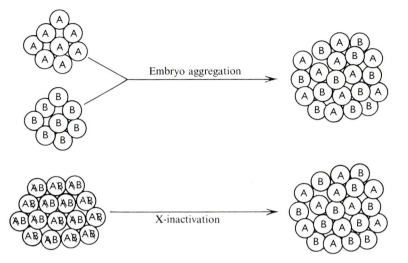

Fig. 32. The similarity between the genetic situation of X-inactivation mosaics and aggregation chimaeras. A and B represent any number of contrasting alleles at loci on the X chromosome, heterozygous in the mosaics and present in different components of the chimaeras.

between them. These will be discussed below, after we have described the attempts that have so far been made to construct and compare aggregation chimaeras and X-inactivation mosaics for a given genetic difference.

The *tabby* locus

Homozygous tabby female or hemizygous tabby male mice have a velvety coat totally lacking in guard hairs, striking abnormalities of the teeth, absence or reduction of many epidermal glands, and no hairs on the tail. Heterozygous tabby females are striped, with narrow light and dark bands of hair arranged transversely down the body, as in 'hair follicle' chimaeras (see Chapter 5). Their glands are more or less normal, and their tails are hairy. Aggregation chimaeras between homozygous or hemizygous tabby embryos and non-tabby embryos have been described by McLaren *et al.* (1973) and by Cattanach *et al.* (1972); the teeth of both groups of animals were analysed by Grüneberg *et al.* (1972).

The seven tabby chimaeras of McLaren *et al.* all showed stripes, qualitatively similar to those seen in tabby heterozygotes, but more irregular in width and number. Three of the seven animals of Cattanach *et al.* showed similar variegation; in two others the coat was largely albino, obscuring any potential tabby markings, in one involving a mottled (Mo^{dp}) component it was typical of a $Mo^{dp}/+$ heterozygote, with only a slight suggestion of tabby banding, and the remaining animal showed no variegation, revealing its chimaerism only through its progeny. The tabby-banded chimaeras ranged from a mostly normal phenotype, with just a few fine stripes on the back, to a phenotype very similar to that of a tabby hemizygote, but with a few fine stripes in the coat and a few hairs on the tail. While the intermediate chimaeras were indistinguishable from heterozygotes, the extreme categories would rarely be found among normal *Ta*/+ females.

Cattanach *et al.* examined individual hairs of their tabby chimaeras and heterozygotes, and found a similar distribution of aberrant 'tabby-type' hairs, normal hairs, and intermediates. Two chimaeras resembled heterozygotes in having a higher proportion of tabby hairs in the dark bands than in the intermediate light areas; the third showed an unexpected and unexplained predominance (99 %) of tabby hairs in the light areas.

In a detailed study of another X chromosome marker, *Greasy* (*Gs*), Dunn (1972) observed that aggregation chimaeras resembled *Gs*/+ heterozygotes both at the level of gross variegation of the coat and with respect to the structure of individual hairs.

Other features of tabby chimaeras and heterozygotes have been examined (McLaren *et al.*, 1973). In both, the glandulae tarsales (glands opening onto the eyelids) were reduced and variable in number, and the papilla vallata on the tongue showed the tabby condition in some individuals and the normal in others. For the plicae digitales (skin folds on the digits), however, the chimaeras were nearer the tabby phenotype than were the heterozygotes. Large differences were also seen in the arrangement of the tail rings (Fig. 33). Normal mice showed a regular pattern of scales, hairs and pigment distribution; heterozygotes showed a different but equally regular pattern. Tabby homozygotes or hemizygotes, on the other hand, showed little or no indication of pattern. The chimaeras differed strikingly from the heterozygotes: the overall banding pattern was much more irregular, consisting of a patchwork of 'tabby' and 'normal' areas, and in addition some 'heterozygous' areas.

The molars of tabby chimaeras were analysed by Grüneberg *et al.* (1972). *Ta*/+ heterozygotes are either indistinguishable in their dentition from the wild-type (35 %) or show a mixture of wild-type and tabby features (65 %). Of thirteen mice seen to be chimaeras from their coats, four had molars similar or identical to the wild-type condition, and three were correspondingly close to the tabby condition. Six were obviously chimaeric with respect to their dentition, with some wild-type and some tabby features. Individual teeth were in general either wild-type or tabby in phenotype, rather than mixed. The comparison between heterozygotes and chimaeras thus showed qualitative agreement, but with a striking shift towards tabby expression (Table 10) in the chimaeras.

Fig. 33. Whole mounts of mouse tail skin. (a) wild-type (+/+), showing regular bands of pigment and hair; (b) heterozygous tabby mosaic (Ta/+) with less pigment but bands still regular; (c) hemizygous tabby (Ta/−), with no regular bands, hairs sparse and abnormal; (d) chimaera (Ta/−↔+/+), showing patches of bald band-free skin alternating with patches of pigment and hairs. From McLaren *et al.* (1973).

TABLE 10. *The percentage of animals showing mutant expression, either partial ('Mixed') or complete ('Tabby'), in respect of their dental morphology*

Genotype	No. of mice	% showing dental morphology of type:		
		Wild-type	Mixed	Tabby
$cr/+$	87	43.7	56.3	0.0
$Ta/+$	121	34.7	65.3	0.0
$Ta \leftrightarrow +$	13	30.8	46.2	23.1

$Ta/+$ animals are heterozygous for *tabby*, an X-linked marker, with the two alleles active in different cells; $cr/+$ animals are heterozygous for *crinkled*, an autosomal mimic of *tabby*, with both alleles active in the same cell; $Ta \leftrightarrow +$ animals are chimaeras, with two distinct cell populations, and a larger patch size than $Ta/+$ heterozygotes.

The albino locus

Mice heterozygous for Cattanach's translocation and carrying the mutant *albino* or *pink-eye* genes on their autosomes are termed 'flecked'. They show a coat colour pattern (Cattanach, 1974) of broad irregular transverse bands similar to that seen in albino↔non-albino (Fig. 34) and other 'melanocyte' chimaeras (see Chapter 5). Cattanach *et al.* (1972) have detected mixed hairs both in 'flecked' females and in albino↔non-albino chimaeras, though in 'flecked' animals the mixed hairs occurred equally in albino and coloured areas, while in the chimaeras they were virtually confined to the coloured areas. In both the mixed hairs proved to be 'polarized', with more pigment near the tip than near the base. The distribution of pigment remained unchanged with age in the chimaeras, but in 'flecked' mice the albino areas became increasingly pigmented as the animals aged, with each hair cycle giving hairs with more pigment than before. Evidence suggests that X chromosome inactivation gradually breaks down with age, permitting activity of the + allele in the rearranged X.

Comparisons between 'flecked' mice and coloured↔non-coloured chimaeras have been made with respect to the eye by Deol & Whitten (1972*a,b*) and by West (1976*a*), and with respect to the ear by Deol & Whitten (1972*b*). In sections of the pigmented retina, patches of pigmented and unpigmented cells can be seen both in 'flecked' mice and in chimaeras. The similarity between the appearance of the retina in the two groups was first commented on by Tarkowski & Cattanach (personal communication, cited by McLaren, 1969). However, the proportions of the two components varied very much more in chimaeras than in 'flecked' heterozygotes, both between animals, between eyes of the same animal, and between sections of the same eye

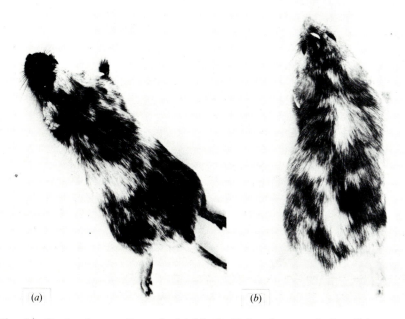

Fig. 34. Coat colour patterns in (*a*) 'flecked' female, mosaic for albino, and (*b*) albino↔coloured chimaera.

(Deol & Whitten, 1972*a*). When the numbers of discrete pigmented and non-pigmented patches were counted, the mean proved to be significantly lower in chimaeras (20 versus 62), and the variance higher.

The distribution of pigmented and non-pigmented cells in the pigmented retina of chimaeras and 'flecked' females has been pursued further by West (1976*a*). He has confirmed the greater variation shown by chimaeras in the proportion of pigmented cells in the retina, and has pointed out (see Chapter 10) that this in itself would lead to a larger mean patch size, and hence a lower mean patch number, in chimaeras. When allowance was made for this effect, the estimated mean number of clones per retina at different ages from 12 days' gestation to adulthood turned out to be closely similar in 'flecked' females and in coloured↔non-coloured chimaeras made between Q-strain embryos. The larger patch size in chimaeras of another strain combination (Table 9) may reflect a tendency for cells of like genotype to stick together.

The distribution of pigment cells with a different developmental origin, namely the migratory melanocytes that colonize the inner ear and the iris, ciliary body and choroid of the eye, has been examined by Deol & Whitten (1972*b*). In chimaeras very large areas were colonized by cells of one component only; in 'flecked' females the albino and pigmented cells formed a much finer intermixture. To what extent this difference reflects greater variation among the chimaeras in the proportion of pigmented cells (see Chapter 10) is not clear from the published report.

Resemblances between mosaics and chimaeras

At the qualitative level, there is little difference in the appearance of X-inactivation mosaics and the corresponding chimaeras. This settles the question, raised originally by Grüneberg (1966), as to whether orderly patterns of the sort seen in X-inactivation mosaics (for example, striping) could be generated from an initially random mixture of two cell populations. Clearly they can.

The fact that orderly patterns can be generated by two populations of cells does not imply that the patterns need be clonally related to the two populations. As we have seen in Chapter 5, the 'melanocyte' patterns characterizing, for example, 'flecked' X-inactivation mosaics and albino↔ non-albino chimaeras are almost certainly clonal, in that the white patches represent areas colonized by 'albino' melanocytes, while the dark areas contain mainly 'non-albino' melanocytes. The mechanisms involved must include clonal proliferation and, at least in the case of melanocyte patterns, ordered cell movement, with a restricted amount of cell mingling. In other situations the pattern may be generated by an interaction between the two alleles, but need not bear any direct spatial relation to their distribution. For example, the pattern of tail rings in tabby heterozygotes (McLaren *et al.*, 1973) is unlikely to reflect the distribution of the underlying cell populations directly, but represents a variation of the pattern seen in the homozygous wild-type animal. The tails of tabby chimaeras are different again, with discrete 'tabby', 'wild-type' and 'heterozygous' areas arranged in no regular pattern. Whether 'hair follicle' banding patterns, such as are seen both in tabby heterozygotes and in agouti↔non-agouti chimaeras, are due to interaction between the two cell populations, or whether they reflect cell lineage, is discussed in Chapter 5.

Grüneberg *et al.* (1972) have pointed out that evidence of interaction between the two cell populations in chimaeras involving X-linked markers throws no light on whether the equivalent interaction between alleles in the corresponding X-inactivation mosaic is intercellular or intracellular. The dental phenotype is qualitatively similar in $Ta↔+$ chimaeras, where the interaction is undoubtedly between cells, and in heterozygotes $(+/cr)$ for *crinkled*, an autosomal mimic of *tabby*, where the two alleles are presumably interacting within the cells to produce the same pattern. The same phenotype in *tabby* heterozygotes $(Ta/+)$ could therefore arise from either inter- or intracellular interaction (Fig. 35). The demonstration that X-inactivation heterozygotes resemble the corresponding chimaeras cannot therefore on its

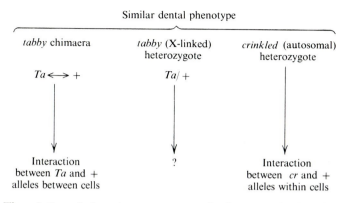

Fig. 35. The relation of phenotype to genotype in three genetic situations, to show that the existence of two cell populations in heterozygotes for X-linked markers cannot be inferred from their phenotypic similarity to chimaeras for the same markers.

own be used to refute Grüneberg's (1967) 'complemental X' hypothesis. However, evidence that two populations of cells exist in X-inactivation heterozygotes is now available from other sources; in addition, the quantitative differences between the dental phenotype in $Ta\leftrightarrow+$ chimaeras, $Ta/+$ heterozygotes and $+/cr$ heterozygotes (Table 10) are most easily explained on the assumption that $Ta/+$ heterozygotes are a mosaic of Ta and $+$ cells, with a patch size smaller than that occurring in $Ta\leftrightarrow+$ chimaeras (see below).

Differences between mosaics and chimaeras

Despite their obvious similarities, chimaeras and X-inactivation mosaics show certain consistent differences from one another, attributable presumably to their different developmental origins. Fig. 36 sets out some possible causal relationships.

In every system so far examined, chimaeras have proved to be more variable than X-inactivation heterozygotes in the relative proportions in which the two components are represented. Heterozygotes contain more or less equal numbers of the two cell populations; in a particular tissue, chimaeras may vary from 100 % of the one type to 100 % of the other. Even considering the composition of the whole body, chimaeras vary very widely. This variation must reflect in part the different numbers of cells of the two components that initially contributed to the genesis of the embryo proper, and in part the results of subsequent cell selection. Chimaeras also differ from X-inactivation heterozygotes in having a larger and more variable patch size, manifest in the coat,

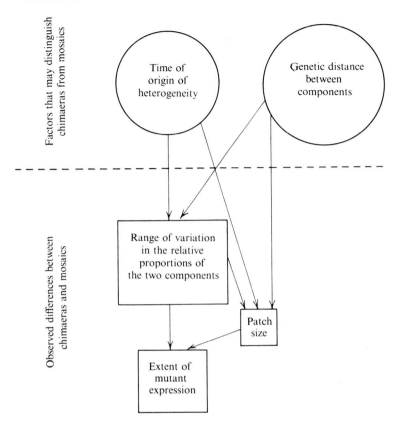

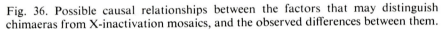

Fig. 36. Possible causal relationships between the factors that may distinguish chimaeras from X-inactivation mosaics, and the observed differences between them.

the tail, the eyes and the teeth, and in showing a stronger tendency towards mutant expression.

This last difference seems to hold for heterozygotes in general, that is for animals heterozygous not only for X-linked markers but also for autosomal genes. More of the tabby phenotype is seen in the tail and teeth of tabby↔ normal chimaeras than of tabby heterozygotes; similarly, the xiphisterna of short-ear↔normal chimaeras show more of the mutant phenotype than do those of short-ear heterozygotes (Chapter 6). The greater range of variation in proportions will in itself mean that some chimaeras will be formed largely from the mutant component, and hence will show the mutant phenotype. However, an equal number should be unbalanced in the opposite direction, being formed largely from the wild-type component. How therefore can the observed predominance of mutant expression be explained?

With X-linked markers, the difference could stem from the larger patch size observed in chimaeras, provided that the character is not cell-autonomous but depends at least partly on cell-to-cell diffusion of a gene product. With small patches, the diffusible product of the normal allele may reach almost all the cells in the tissue; very little expression of the recessive mutant phenotype would then be seen. As patch size became larger relative to the 'effective radius of diffusion' of the gene product, so more of the interior of each mutant patch would remain unaffected, and more mutant expression would be expected. With autosomal markers, the effect should be even more extreme, as the interaction between alleles in the heterozygote would occur within a single cell. Some relevant data on chimaeras and heterozygotes for X-linked and autosomal markers are given in Table 11.

Thus the greater degree of mutant expression in chimaeras may merely reflect the larger patch size. We should therefore seek next an explanation of the different patch sizes characteristic of chimaeras and X-inactivation mosaics.

In their origin, X-inactivation mosaics and chimaeras differ in two major respects:

(1) Developmentally, the heterogeneity in a chimaera arises earlier than in a mosaic. Aggregation chimaeras are mostly made in mid-cleavage, $2\frac{1}{2}$ days *p.c.*; X-inactivation does not occur before $4\frac{1}{2}$ days *p.c.* in the mouse, and possibly not till a couple of days later. There is also the possibility that the two components in chimaeras may differ slightly in their developmental stage, but the implications of this for subsequent development are not known.

(2) Genetically, the two cell populations in a mosaic differ only at those loci for which the X chromosomes are heterozygous. The genetic difference between the two components of a chimaera is usually (though not always) much greater.

Both these differences are likely to contribute to the greater variance shown among chimaeras in the relative proportions of the two components. Both could also independently account for the larger and more variable patch size characteristic of chimaeras, while the greater variance in the proportions of the two components would in itself be expected to result in a larger mean patch size (see Chapter 10).

The greater the genetic distance between two components, the more likely is it that forces of cell selection and differential proliferation will favour one at the expense of the other. With highly inbred and therefore genetically homogeneous components, the advantage is likely always to be in the same direction; but with random-bred material, as has been used for much of the chimaera work (including both the tabby studies), the advantage could be in favour of either component. This would increase the variation between chimaeras. Patch size might be affected directly by the genetic relationship between the components if some cell recognition system existed such that genetically similar cells tended to stick together. No such tendency has been demonstrated directly, though organ-specific differential adhesion of dis-aggregated cells in culture is well established. The results from chimaearas of one strain combination, (C57BL × C3H)F$_1$↔multiple recessive, suggest that some genotype-specific cell adhesion may be occurring (see Chapter 10). In contrast, a series of chimaeras between congenic strains (CBA-T6↔CBA-p; Mystkowska & Tarkowski, 1968), in which the components are very close to one another genetically, closer than in most X-inactivation mosaics, showed evidence of a rather small patch size, both in the coat and in the spermatogenic tubules (see Chapter 4).

The time of origin of heterogeneity could also affect both patch size and the variation in proportions of the two components. If development encom-passes a succession of sampling events (Chapter 10), the earlier in this succes-sion the heterogeneity arises, the greater will be the variation among animals and among organs. In particular, the derivation of the foetus from a rela-tively small number of cells in the inner cell mass of the blastocyst must be responsible for much of the variation seen among chimaeras. X-chromosome inactivation has not occurred at this stage of development, so this cause of variation will be absent in X-inactivation mosaics.

As for patch size, it is clear that in a situation of largely clonal development with restricted cell mingling, the later heterogeneity arises, the smaller will be the final patch size. This argument has been used by Deol & Whitten (1972a, b) and by McLaren *et al.* (1973) to suggest that X-inactivation in some tissues may not take place until a relatively late stage of development.

Nesbitt (1974) has claimed that the larger patch size reported by Deol & Whitten (1972a) in the pigmented retinas of chimaeras than of X-inactivation mosaics would be consistent with X-inactivation occurring at about the time of implantation. She argues that provided cell mingling occurs to the same extent in both groups and is restricted, so that the distribution of the two cell populations never becomes random, and provided that the disruptive effect of cell migration on clone size is independent of initial clone size, the ratio of the mean patch sizes in the adult retinas should remain unchanged from the time of X-inactivation to the adult. (In fact Lewis (1973) has shown that the effect of cell migration is not independent of initial clone size, but disrupts small clones to a greater extent than large clones. This will tend to increase the patch size ratio of chimaeras relative to mosaics as development proceeds.) Since in the adults examined by Deol & Whitten (1972a), the mean patch size in the pigmented retinas was ninefold greater in chimaeras than in mosaics, Nesbitt infers that at the time of X-inactivation the chimaeric patches, and hence also clones, would have been not more than nine times greater than those in mosaics, i.e. each chimaera clone should have contained no more than nine cells. It is known that X-inactivation does not take place before the blastocyst stage (Gardner & Lyon, 1971). A chimaeric embryo, at least up to the blastocyst stage, is in fact made up of only two patches, one corresponding to each component (Garner & McLaren, 1974): each consists of a single clone, containing for example thirty-two cells in a 64-cell blastocyst. The clones will presumably be partially disrupted by cell movement during downgrowth of the egg cylinder, until they contain no more than nine cells each. If X-inactivation occurs at this point in time, producing a clone size of one in the mosaics, the ratio of 9:1 will have been realized. However, this argument depends on restricted (or no) cell mingling throughout later development, and West (1976a) has shown that this condition is not fulfilled for the pigmented retina: the distribution of the two components on the 13th day of gestation is essentially random both in chimaeras and in mosaics.

Can the distribution of the two cell populations in chimaeras and mosaics in fact throw any light on the time of X-inactivation? West (1976a) observed in the pigmented retina no difference in mean clone size between mosaics and some chimaeras, when allowance was made for the effect on patch size of variations in the proportions of pigmented cells. In other chimaeras clone size was larger, perhaps owing to differential cell adhesion in that particular strain combination. West points out that the difference in patch size observed by Deol & Whitten (1972a) could similarly reflect a difference between mosaics and chimaeras either in the proportions of pigmented cells in the retina, or in differential adhesion. To establish that a difference in patch size resulted from a difference in the time of origin of the heterogeneity, it would thus be necessary first to make allowance for possible differences in the relative proportions of the two components, and then to demonstrate that cell mingling had occurred to the same extent in both groups, and had never resulted in a random arrangement of the component types. These conditions are unlikely to be fulfilled: it therefore seems that evidence on the time of X-inactivation will need to be sought elsewhere.

10

Distribution of cell populations

Chimaeras, by definition, contain two genetically distinct cell populations. If markers are used that act locally, so that they are expressed within the cells that carry them, the spatial distribution of these populations can be investigated. This in turn may yield information on the developmental history of the animal, how it came to be the way it is.

Markers may be visible in the intact animal with the naked eye, as with coat colour markers, or in the intact tissue after histochemical treatment, as with β-glucuronidase in liver cells. More often, they may be detected only after disruption of the tissue, either in individual dividing cells (chromosome markers) or in chunks of tissue (enzyme markers). The scarcity and inadequacy of existing cell markers in mammals has already been emphasized (Chapter 2).

In *Drosophila*, the distribution of cell populations in chimaeras and mosaics displays developmental history in a very direct way. If the first cleavage division gives rise to two non-identical cells, one half of the adult fly, say the left half, or the front, will be of the one type, and the other half will be of the other type. If a genetically distinct cell arises or is introduced at a later stage of development, its progeny forms a cohesive, spatially contiguous patch of cells in the adult fly. This makes 'fate maps' of the early embryo relatively easy to construct. In mammals, on the other hand, cell mingling and cell movement, cell selection and cell death are all rife, making the distribution of cell populations in the adult far less easy to interpret in developmental terms.

Clones and patches

The progeny of a single cell is termed a clone. In this sense any normal individual is a clone, derived from a single zygote; a chimaera formed by aggregating two embryos contains two such clones. At any time during development, each cell in the embryo is the progenitor of a smaller clone; the later the stage, the smaller is the clone. During periods when cell proliferation predominates over cell movement, so that most or all of the members of a clone remain coherent, in the sense of spatially contiguous, development will be termed 'clonal'. Nesbitt (1974) has distinguished between 'strict coherent clonal growth', in which mitotic daughter cells invariably remain adjacent, and 'limited coherent clonal growth', in which cell migration occurs and disrupts clones to a degree, but not so much that cell distribution within the embryo is randomized, as it would be in 'nonclonal growth'. The value of chimaeras in allowing periods of clonal development to be identified has been emphasized by Mintz (e.g. 1972*b*, *c*).

The presence of spatially contiguous cells of one component type ('patches') in a chimaera cannot in itself be taken as evidence for clonal development. Even if development were entirely non-clonal, so that the two types of cell were arranged at random, we would expect to see 'patches', in the sense of aggregates of cells of one type totally surrounded by cells of the other, provided that the two types were phenotypically distinguishable. Curtis (1967) pointed out that, 'in a randomly packed aggregate composed of equal numbers of cells of two types', the majority of cells of one type 'are in contact with cells of the same type through a network structure'. Curtis was considering three-dimensional arrays of cells; similar arguments apply to two-dimensional arrays, while for the one-dimensional case (e.g. a section through a two-dimensional sheet of cells) the average patch size can be calculated algebraically as $1/(1-p)$ where p is the proportion of one cell type in the population. For the two- and three-dimensional case the average patch size in a random aggregate has not yet been calculated algebraically, but may be estimated from computer models. When the observed patch size in a chimaera significantly exceeds the patch size estimated for a random aggregate, one may infer that some degree of clonal development has taken place; the extent of the difference reflects the number of cell generations for which development has been clonal.

Some confusion has arisen in the past because the number of patches in an adult chimaera has been taken to represent the number of clones. The relation between clones and patches has been clarified, with reference to computer models, by West (1975) for one- and two-dimensional situations, by Ransom, Hill & Kacser (1975) for two-dimensional situations, and by W. K. Whitten (personal communication) for three-dimensional situations. The one-dimensional case was discussed, with reference to coat colour patterns, by Wolpert & Gingell (1970).

The average patch size in a random array depends critically on the relative proportions of the two cell types in the population. In a one-dimensional array the average patch size is minimal, and hence the total number of patches is maximal, when the two types are present in equal proportions (i.e. $p = 0.5$, $1/(1-p) = 2$). For a two-dimensional, and still more for a three-dimensional array, as the proportions depart from equality the more numerous type increasingly aggregates into a branching network, ultimately forming a single large branching patch, while the minority type forms large numbers of very small patches; the total number of patches is then at a minimum when $p = 0.5$ because both populations show some aggregation of patches.

In a large two-dimensional array with $p = 0.5$, the average patch size in a random arrangement is about twenty-five. This means that if the observed *patch* size in an appropriate chimaeric tissue such as the pigmented retina was 100, with pigmented and non-pigmented cells present in equal proportions, the estimated *clone* size would be four (i.e. $100 \div 25$), so that on a simple model development would have been clonal for the last two cell generations only (West, 1975).

An alternative method of estimating clone size is to look at the distribution of patch sizes: the smallest patch should represent a single clone derived from a single cell, the next smallest patch should be twice as big, the next three times as big, and so on. This assumes homogeneity of development (e.g. rate of cell division) between the two cell populations and between different parts of the tissue. Deviations from these conditions will be reflected in the distribution of patch sizes. In practice overlaps between size classes will occur, since cells vary in their rate of cell division; a relationship between the variation in doubling time and that of patch size is given in Ransom *et al.* (1975).

The clonal analysis of development in chimaeras would also be affected by any tendency for genetically like cells to stick together. No such differential adhesion has yet been demonstrated directly, *in vivo* or *in vitro*. Circumstantial evidence suggests, however, that in the series of chimaeras made between $(C57BL \times C3H)F_1$ embryos and embryos of a multiple recessive strain (McLaren & Bowman, 1969), the two component cell types mingle less freely than in other combinations. The most convincing demonstration comes from the analysis by West (1976*a*) of patch size in the pigmented retina, in which chimaeras of this strain combination proved to have significantly larger patches than other chimaeras (Table 9). In the coat, the transverse bands reflecting melanocyte migration tended to be particularly distinct (see Chapter 5). The infrequency of mixed progenies from chimaeras of this strain combination suggests that the germ cells colonizing the gonads all belonged either to one component type or to the other. When mixed progenies did occur, the proportions of the two types of progeny in successive litters were strikingly heterogeneous, differing significantly in this respect from chimaeras in which the two components were more closely related (McLaren, 1975*a*).

The degree of mixing of the two cell populations could have a critical effect on development in the special case of sex chimaerism. Tarkowski (1969) has pointed out that the more thorough the mixing of XX and XY cells in the genital ridge, the less likely might be the formation of an ovotestis. However, the strain combination described above, that tended to show a relatively large patch size, did not have a high incidence of hermaphroditism.

Visible patchiness in the adult

Coat colour patterns have been discussed in detail in Chapter 5. For those that depend on melanoblast distribution (e.g. in albino↔coloured chimaeras), Mintz (1967a, 1970b, 1971b) has inferred from the maximum number of transverse stripes seen in different regions of the body that the number of melanoblast clones originating from each side of the neural crest is seventeen. Stripes of course represent patches, in the sense used above; Mintz' argument is equivalent to the method of estimating clone size by taking the smallest patch as representing a single clone. The estimate should be taken as referring to the number of discrete clones in the neural crest at the time that melanoblast migration begins, rather than the number at the time of 'determination' or 'commitment'.

Transverse striping is equivalent to a one-dimensional array; the average number of clones per patch is therefore expected, from the $1/(1-p)$ relation, to be two in a balanced ($p = 0.5$) chimaera. Hence seventeen clones per side (whether precisely or as an average) would be seen as an average of $8\frac{1}{2}$ patches (i.e. stripes) per side. The average number of stripes in our data presented in Fig. 20 is 8.14: this fits well with expectation, especially since not all the animals examined were balanced, and as p diverges from 0.5, so the expected number of stripes declines.

A further complication of cell distribution is illustrated in Fig. 37, which shows the distribution of pigment along the length of the body in the same series of chimaeras (West & McLaren, 1976). The paler, recessive component was represented more strongly in the anterior than in the posterior part of the body. Since the migration of melanoblasts outwards from the neural crest begins in the anterior part of the crest and spreads backwards, this observation suggests that the two components may differ in the timing and pattern of migration of their melanoblasts, and that the differences behave autonomously in the chimaeras. For instance, recessive melanoblasts may begin their migration earlier in development, and so be at a selective advantage in the anterior regions; but if the backward spread of migration is slower for recessive cells, they will be at a disadvantage in the more posterior part of the body. A similar explanation, namely a temporal shift of strain-specific selective advantage, has been suggested by Moore & Mintz (1972) for the antero-posterior gradient in the proportion of C57BL cells that they observed (Fig. 26) in the vertebrae of C57BL↔C3H chimaeras (Chapter 6).

With coat colour genes acting through the hair follicles (e.g. *agouti*, *fuzzy*), a much finer pattern of transverse striping is seen in chimaeras. A similar pattern is characteristic of females heterozygous for X-linked genes acting through the follicles (e.g. *tabby*). Mintz (1970b) has postulated that these narrow stripes correspond to somites (or to hypothetical 'transient' somites in the head), and hence represent mesodermal clones. She estimated that

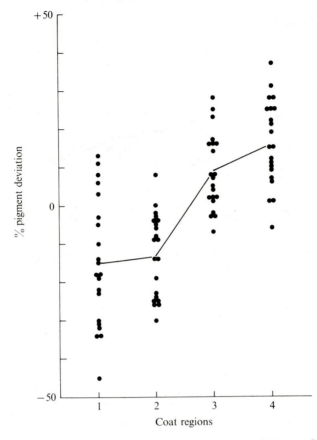

Fig. 37. The distribution of pigmented hairs in (C57BL × C3H)F$_1$↔recessive chimaeras. The dorsal coat was divided into four approximately equal regions, of which the head was region 1 and region 4 was the most posterior. Patches were identified, the degree of pigmentation of each patch was assessed subjectively, using a 5-point scale, and the area of the patches measured with a planimeter. The pigmentation of each region was expressed in terms of its percentage deviation from the level of pigmentation of the entire dorsal coat. From West & McLaren (1976).

there should be 170 such clones, eighty-five per side, each corresponding to a somite. If this were the true number of clones, the expected number of stripes would be only half as great, even in a balanced chimaera. The estimate was based partly on extrapolation, assuming each clone corresponded to one somite, and partly on direct observation of fuzzy↔non-fuzzy chimaeras (Mintz, 1970*b*); however the counts or measurements of stripes in these animals have not yet been published.

Visible patchiness does not always reflect clonal origin, but may spring from environmental differences: for example, the black ears and feet and tail tip of a Siamese cat or Himalayan rabbit give no information on clonal development, but are a response to the lower body temperature at the extremities. A striped pattern could reflect rhythmical changes in the microenvironment of the epidermis, e.g. oscillations in the concentration of some chemical substance. As pointed out in Chapter 5, the evidence in favour of a clonal explanation of the pattern is very strong in the case of the broad stripes of the melanocyte chimaera, less strong for the narrower stripes of the hair follicle chimaera. Accurate information on the statistical distribution of numbers or widths of stripes would enable the clonal theory to be tested; unfortunately the narrow stripes of the tabby heterozygote or fuzzy↔non-fuzzy chimaera are hard to count or measure.

The distribution of pigmented and unpigmented cells may conveniently be studied in the pigmented retina of chimaeras carrying appropriate colour markers, or in 'flecked' females heterozygous for Cattanach's X–autosome translocation (see Chapter 9). In the neural retina, the gene for *retinal degeneration* (*rd*) may be used as a marker, as areas of retinal degeneration can be seen on sections. Mintz (1971*c*) reckons that the pigmented and the neural retina each consist of sectors radiating from a common origin. The sectors in the two layers bear no relation to one another, at least in the adult. Mintz & Sanyal (1970) estimated that in the neural retina, the sectors number ten per eye, and postulated that they represent clones. The term 'clone' is used by Mintz in a restricted sense, to mean 'the mitotic progeny of one cell in which a specific constellation of gene loci first became active or derepressed, and has remained active (or mobilizable) as a cell heredity'.

Using a rather different approach, West (1976*a*) has measured patch size in sections of the pigmented retina of pigmented↔unpigmented chimaeras and in 'flecked' mice (see Chapter 5), at different ages both before and after birth (see Fig. 38). At $12\frac{1}{2}$ days' gestation, when pigment first develops, the patch size is consistent with a random arrangement of the pigmented and unpigmented cells, i.e. mean clone size is not much more than one cell. In the adult, patches are larger by an amount that suggests a clone size of four to eight cells, i.e. two or three cell generations of coherent clonal growth. The failure to detect any clonal pattern at $12\frac{1}{2}$ days suggests that the 'radiating sectors' observed by Mintz may arise during later development, presumably by directionally oriented cell division. A similar approach to the neural retinal was unsuccessful (West, 1976*d*): the existence in chimaeras of areas intermediate between the normal and the 'retinal degeneration' phenotype complicated the measurement of patch size, and suggested either that *rd* does not act entirely autonomously, within the cell, or that a considerable amount of migration had occurred.

The vertebral column may also exhibit visible patchiness. From a sophisticated shape analysis of the vertebrae in C3H↔C57BL chimaeras, Moore &

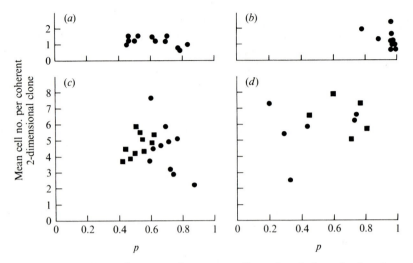

Fig. 38. Mean cell number per coherent two-dimensional clone in the pigmented retinal epithelium, estimated from longitudinal (●) and latitudinal (■) sections of eyes of 'flecked' mosaics and unpigmented-Q↔pigmented-Q chimaeras at $12\frac{1}{2}$ days' gestation and in the adult. (*a*) mosaics at $12\frac{1}{2}$ days' gestation; (*b*) chimaeras at $12\frac{1}{2}$ days' gestation; (*c*) adult mosaics; (*d*) adult chimaeras. Note that clone size is independent of *p*, the proportion of pigmented cells in the retina. From West (1976*a*).

Mintz (1972) concluded that at least four cells or clones must contribute to each vertebra, making a total of 240 for the entire vertebral column. The argument rests on the assumption that the C3H and C57BL cells act locally, within the individual vertebra.

The liver offers further opportunities of visualizing cell distribution in chimaeras, as strain differences exist in the activity of hepatic β-glucuronidase; these can be detected histochemically (Fig. 7) and appear to be cell-autonomous. Patches staining red and green, corresponding to high- and low-activity cells, were first illustrated by Condamine *et al.* (1971); the first attempts to estimate the size of such patches have been made by West (1976*c*). Cells were counted in two-dimensional sections of the liver, and in one-dimensional transects across sections. The estimates of three-dimensional patch size agreed well with one another, and suggested that the adult liver contains 10–20 million spatially contiguous clones, with ten to thirty-four cells per clone.

A much lower number was derived by Wegmann (1970), using a more indirect technique. β-Glucuronidase activity was compared in samples of liver from chimaeras and from F_1 hybrids between high- and low-activity strains. The chimaera samples proved to be much more variable than those from F_1

hybrids, as would be expected if the chimaera livers were made up of patches of high- and low-activity tissues, of a size comparable with that of the samples. The fewer and larger the patches, the more variable would be the chimaera samples. Using a binomial approach, Wegmann derived a figure of 17.2 per sample, say 340 for the liver as a whole, which must be interpreted as the number of patches rather than the number of clones (cf. Nesbitt & Gartler, 1971). In a three-dimensional system, this could correspond to as many as 1 400 000 clones (W.K. Whitten, personal communication). However, the discrepancy between West's and Wegmann's estimates of clone number is still large.

The statistical approach

Often it is impossible to see patches of the two component types in a chimaera, yet appropriate techniques (isozyme determinations, chromosome preparations) will reveal whether both components or only one are present in any tissue or organ. The frequency with which only one component is represented in a given tissue can yield information on the number of cells from which that tissue is derived. If it originates from a single cell, no more than one component can ever be present. Provided no complicating factors intervene, one component only will be present in 50% of cases if the tissue originates from two cells, and in $1/(2^{n-1})$ cases if it originates from n cells.

In a large series of C3H$\leftrightarrow$C57BL aggregations, Mintz (1970b) observed that 70% of aggregants were overt chimaeras, with 71% of these showing chimaerism in the hair follicles and only 44% in the liver. Taking into account the sensitivity of the markers, this allowed her to infer very tentatively a three-cell 'clonal origin' for the embryo as a whole, and also for the hair follicles, and a two-cell origin for the liver. The 'compartments' were envisaged as 'intermediate, not final, stages of genetic determination', with the 'final derivation' arising later, in larger numbers of cells, e.g. 150–200 for the hair follicles. A similar binomial argument applied to X-inactivation mosaics led Gandini *et al.* (1968) to infer that the haematopoietic system derived from a pool of eight cells or less at the time of X-inactivation, and Tettenborn, Dofuku & Ohno (1971) to postulate 'four or five embryonic progenitor cells' for the kidney proximal tubule cells. Wegmann's (1970) studies on the liver, described earlier, represent a similar approach, though based on a quantitative rather than an all-or-none criterion.

Two types of question may be asked of this type of analysis: (*a*) What do the estimates mean? (*b*) Are the assumptions justified?

Much confusion has arisen from phrases such as 'The number of cells from which the tissue was derived' or 'The number of progenitor cells' or 'cells committed to serve as progenitors' or 'The number of cells set aside for the development of the tissue' or 'The initial pool size for the tissue'. In its immediate past, every tissue containing n cells has been derived from $n/2$ cells one generation ago. More remotely, every tissue has been derived from a single cell, the zygote, in an ordinary animal or a mosaic, and from two zygotes in a chimaera. Attempts have been made to specify more precisely the stage in development to which the estimate applies: e.g. the stage 'at which the tissues first become differentiated' or 'at which tissue-specific genes are first switched on' or 'at commitment' or 'at inactivation'.

The problem has been discussed at length by McLaren (1972b) and by Lewis, Summerbell & Wolpert (1972). It is concluded that the approach is only meaningful if applied to a system in which a period of random cell mingling of indefinite duration is succeeded by a period in which the tissue is formed by coherent clonal growth. Subsequent to this moment of *allocation*, cell mingling may take place within the tissue primordium, but no significant movement of cells occurs between the primordium and its neighbours. The binomially derived estimate of initial cell number relates either to the point at which allocation occurs, or to the point at which heterogeneity arises (mid-cleavage for aggregation chimaeras, after implantation for X-inactivation mosaics), whichever happens later in development. There is no *a priori* reason why allocation should coincide with biochemical or morphological differentiation, or with the activation of tissue-specific genes.

One major assumption required for any valid inference to be based on binomial reasoning is that the two cell types should be distributed randomly at the time of sampling. Thus in the derivation of the embryo proper, the three 'progenitor' cells would presumably be located in the inner cell mass of the blastocyst. Yet Garner & McLaren (1974) found that, far from being arranged randomly, the cells of a chimaeric blastocyst formed two coherent clones, one of each component type. The assumption of random distribution of cells has not been tested for any of the later stages of development at which sampling events are likely to occur.

The other assumption underlying the binomial approach is that there is no differential cell death or proliferation of one component at the expense of the other. In X-inactivation mosaics, where the two components are genetically identical except for their active X chromosome, such an assumption may hold; in chimaeras, it is often demonstrably false. The estimate of three 'progenitor' cells for the embryo proper is based on the premise that 25 % of aggregated embryo-pairs show no evidence of overt chimaerism. In fact the values of this proportion so far reported range from under 20 % to over 60 % (Mullen & Whitten, 1971); it will include aggregants from which one component has been lost through technical hazards, such as developmental retardation or defects induced by manipulation or culture, as well as innate differences in mitotic rate or viability. As for differential proliferation, Mintz (1970e) has emphasized how in any particular strain combination, one component or the other is likely to be at a selective advantage in any particular organ. For example, AKR cells combined with those of other strains tend to predominate in all tissues except melanocytes and germ cells, and to constitute the overwhelming majority in lymphomyeloid tissues, whether examined cytogenetically (Ford *et al.*, 1974) or by enzyme determinations and serum allotypes (Barnes, Tuffrey, Drury & Catty, 1974*b*). In C3H↔C57BL chimaeras the two strains tend to have approximately equal representation in the kidney, the C3H component predominates in liver and mammary gland, while the C57BL component appears to be at a selective advantage in erythropoietic tissue and gamma-globulin-producing cells – hence the concept of Sam, the Statistical Allophenic Mouse (Mintz, 1970*e*), each of whose tissues has the most likely composition for the given strain combination. Without prior knowledge of the appropriate Sam's properties, the binomial approach is likely to fall into serious error.

Rapidly dividing tissues should show the effects of cell selection particularly clearly. If cell division continues into postnatal life, as in spermatogenesis and haematopoiesis, a change with age in the balance of the two components may be seen. Evidence for temporal shifts in germ cell and erythrocyte populations is cited in Chapters 4 and 7 respectively.

A more elaborate (though related) statistical approach examines the comparison between the proportions of the two chimaera components in different cell populations within an individual. Correlation analyses are used to estimate whether the two cell populations are independent, or are members of a single, larger population.

The overall picture that emerges from these studies is of successive sampling events, with gradual restriction of developmental potential. Granulocytes and erythrocytes appear to derive from a common pool of stem cells at the time of X-inactivation; lymphocytes derive from different precursors (Gandini *et al.*, 1968), but with a common ancestral cell pool at an earlier stage (Mintz, 1971*b*), producing a correlation between the relative proportions of the two haemoglobin and the two gamma-globulin types in chimaeras (Mintz & Palm, 1969; Wegmann & Gilman, 1970). Within the lymphomyeloid system, cells of the bone marrow, thymus and Peyer's patches seem to belong to one population and spleen and lymph node cells to another (Ford *et al.* 1975*b*); both populations seem to derive from a single cell pool earlier in development (Gornish *et al.*, 1972). The lymphomyeloid system as a whole (Ford *et al.*, 1974) shows some correlation with the melanocyte population, judged by coat colour, which in turn is positively correlated with the expression of the two components in the skeletal system (Grüneberg & McLaren, 1972).

The most comprehensive application of such covariance analysis has been made, not on chimaeras but on X-inactivation mosaics (see Chapter 9), by Nesbitt (1971). In a variety of tissues of mice heterozygous for Cattanach's X–autosome translocation she examined the proportion of cells that had the translocated X chromosome inactive rather than the normal one. Differences between mice in respect of the mean level of mosaicism for the tissues examined (that is covariance within individuals) suggested that a random sample of about twenty-one cells is common to the ancestry of all these tissues. This sampling event Nesbitt equated with X chromosome inactivation. From the independent variation of the different tissues within individuals she deduced that a further sampling event, unique to each tissue type, involves twenty to fifty cells, and speculated that it consists of 'determination of the primordial precursor cells for the tissue'.

This type of analysis raises the same problems of interpretation, and rests on the same rather fragile assumptions, as have been formulated above in relation to the single-tissue binomial approach. The various additional factors modifying the relative representation of the two components in different tissues, such as convergence due to similar cell selection pressures, have been interestingly discussed by Ford *et al.* (1975*b*) and by Nesbitt (1971) with reference to chimaeras and mosaics respectively.

Conclusions

In thinking about cell lineages, one can start either with the early embryo and work forwards, or with the finished product, the adult, and work backwards. If these two approaches were able to meet in the middle, we would have the beginnings of a conceptually satisfying model of cell distribution in development.

Little if any cell mingling takes place during cleavage, up to the blastocyst stage. Garner & McLaren (1974) reached this conclusion after aggregating unlabelled 8-cell embryos with embryos labelled with tritiated thymidine, and examining serial sections of the resulting blastocysts two cell divisions later, by autoradiography. On the other hand, a considerable amount of cell mingling must take place some time during development: one or a few cells injected into the blastocyst can contribute to every adult tissue examined, even to areas as small as individual lymph nodes and segments of gut a few millimetres long (Ford *et al.*, 1975*b*), and the patches in for example the pigmented retina are small.

The only direct information on cell distribution during embryogenesis comes from the rat↔mouse chimaeras of Gardner & Johnson (1973). Patch size in the egg cylinder appeared to be relatively large. This might suggest little cell mingling even during egg cylinder formation; however, most of the chimaerism was observed in the proximal endoderm, a tissue that probably does not contribute to later development. Further, interactions between rat and mouse cells may not be typical of those that occur in normal mouse development or mouse↔mouse chimaeras. Further information on the development of rat↔mouse chimaeras may be found in Gardner (1975*a*), Gardner & Johnson (1975) and Johnson & Gardner (1975).

For later stages of development, the elegant study of Gearhart & Mintz (1972*a*) gives some inkling of the complexities of cell distribution. Aggregation chimaeras were made between embryos distinguished by genetic variation for the enzyme glucose phosphate isomerase (GPI). On the 8th and 9th days of gestation individual somites were dissected out and examined by electrophoresis. Thirty of the thirty-eight somites proved to contain both enzyme variants; the relative proportions of the two components tended to be similar in neighbouring somites. Eye muscles from adult chimaeras, each muscle known to be derived from a single somite, were also examined. Every muscle (33/33) contained both enzyme variants. The myotubes of skeletal muscle originate by cell fusion: any myotube of mixed constitution for a dimeric isozyme such as GPI will express not only the two components but also a band of hybrid enzyme. However, 11/33 of the eye muscles lacked hybrid enzyme.

What can we infer from these results? The simplest form of clonal development that could be envisaged for the somites would be if each somite were derived from a single separate cell at some stage of embryogenesis, so that each somite represented a single non-overlapping clone. This model is ruled out by the finding that each somite contains both components. Alternatively, groups of cells from a randomly ordered array could be allocated to form individual somites. This model too is ruled out, as it would not account for resemblances among neighbouring somites. Evidently some degree of clonal proliferation as well as cell mingling must precede the period of somite allocation: the relative proportions of the two components would then be distributed non-uniformly along the array, and adjacent samples of cells could give rise to somites each containing both components in varying but characteristic proportions. The relative infrequency with which hybrid enzyme was seen in the eye muscles suggests that each sample, destined to give rise to a somite, contains rather few cells and subsequently undergoes rather little internal cell mingling, so that most of the cell fusion involved in myotube formation unites cells of like genetic type.

A somewhat similar picture emerges from West's (1976*a*) study of cell distribution in the pigmented retinas of coloured↔non-coloured chimaeras and X-inactivation mosaics. The first appearance of pigment, on the 13th day of gestation, must be preceded by a period of active cell mingling, as the two types of cell are arranged more or less randomly in the two-dimensional cell sheet, with a clone size not significantly in excess of one cell. In one strain combination the clone size was somewhat larger, raising the possibility of differential cell adhesion. Subsequent expansion of the pigmented retina is brought about mainly by clonal proliferation, with some increase in cell size, and a small amount of cell mingling that produces a small increase in the number of clones.

Thus development can be envisaged as consisting of alternating periods

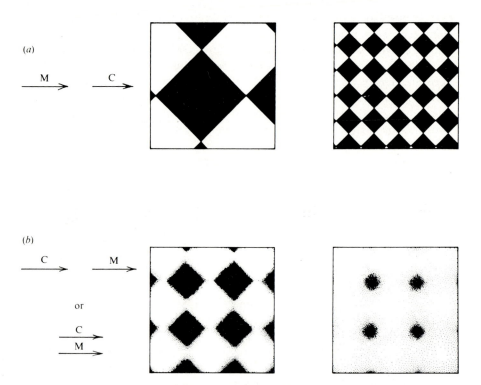

Fig. 39. Diagram of the types of pattern generated by periods of cell mingling (M) and clonal proliferation (C) differing in their duration, ordering and intensity. (*a*) mingling followed by proliferation gives sharp patches, larger or smaller according to the amount of proliferation. (*b*) mingling after or during proliferation gives patches with edges less or more blurred according to the amount of mingling.

in which either cell mingling or clonal proliferation predominates, with successive sampling events allocating groups of cells to particular organs or tissues. If a period of clonal proliferation follows a period of cell mingling, no areas of intermediate phenotype will be seen (provided of course the character is autonomous, i.e. the gene product does not diffuse from cell to cell) and the patch size will depend only on the duration of the period of clonal proliferation (Fig. 39*a*). If, on the other hand, clonal proliferation has preceded cell mingling, areas of intermediate phenotype may be found, alternating with 'single-type' patches, the size of which will depend on the relative lengths of the proliferative and the mingling periods. If the two activities proceed simultaneously (biologically perhaps the most likely situation), or if the character is not fully autonomous, areas of intermediate phenotype may again be seen, with the size of the 'single-type' patches depending on the relative intensities of clonal proliferation and cell mingling, as well as on the initial patch size (Fig. 39*b*; see Lewis, 1973).

Further, in any particular chimaeric strain combination, differential viability or proliferation may lead to cell selection, perhaps varying in intensity or direction from tissue to tissue, and differential cell adhesion may restrict cell mingling. Increased understanding of this system is unlikely to come from further application of statistical techniques based on relatively simple models, but must await the discovery of new cell markers that can be applied to hitherto inaccessible periods of development.

11

Spontaneous chimaeras

Every mammal is an organized patchwork of different kinds of cell. Apart from the phenotypic differences characteristic of 'normal' cellular differentiation, so obvious yet so deeply mysterious, cells differ with respect to their chromosome organization. Normal diploid cells exist side by side with haploid cells in the testis, with polyploid cells in the liver and, in the mouse, with polytene cells in the placenta. In a female mammal, both X chromosomes seem to be functional in germ cells, but only one in somatic cells. In eutherian mammals, the choice between activation of one or other X chromosome is initially made at random, but the distribution of the two resultant cell lines in subsequent development is governed by regularities of cell migration and proliferation.

More rarely, cells become different as a result of some accident. For example, a chromosome may be lost at mitosis, producing an XO cell line in an otherwise XX or XY individual; or non-disjunction may lead to an XO and an XXX or XXY cell line. Somatic mutation or somatic crossing-over can also initiate a genetically distinct cell line, as in sectoring of the iris, or the development of a patch of albino skin.

All individuals showing such heterogeneities are termed mosaics rather than chimaeras because, however many cell lines are present, they all derive from a single zygote. However, individuals are sometimes born, both animal and human, containing two cell lines differing from one another in several genetically determined features. These must have been derived from more than one zygote, and thus constitute spontaneously arising chimaeras, according to the definition given in Chapter 1. If one cell line is identified only in the blood (e.g. Brøgger & Gudersen, 1966), the chimaerism may be secondary, that is it may have arisen after implantation by transfusion of blood cells from the mother or from a twin; if both lines are also present in solid tissues, the individuals are presumed to be primary chimaeras.

Most cases of spontaneous chimaerism that have been detected in Man and other mammals have one cell population with a normal female chromosome complement (XX) and one with a male complement (XY). They often show ambiguous external genitalia, or other anatomical evidence of hermaphroditism. The origin of two such cell lines from a single zygote is unlikely, but not impossible: for example an XXY embryo could undergo two non-dis-

TABLE 11. *Spontaneous primary chimaeras in Man*

Evidence of chimaerism	Cell line markers	No. of contributions demonstrated		Reference
		Maternal	Paternal	
Blood and retinal chimaerism Hermaphroditism	XX/XY, MN, S, Rh	1	2	Gartler *et al.* (1962)
Blood and fibroblast chimaerism Hermaphroditism	XX/XY Haptoglobin	1	2	De Grouchy *et al.* (1964)
Blood and skin chimaerism	XX/XY, ABO, S, Jk, sickling, skin colour	2	2	Zuelzer *et al.* (1964)
Blood chimaerism Hypospadias	XX/XY, ABO, Jk	1	2	Myhre *et al.* (1965)
Blood and fibroblast chimaerism Hermaphroditism	XX/XY, ABO, acid phosphatase, phosphoglucomu- tase	2	2	Ferguson-Smith *et al.* (1966), cited by McLaren (1969)
Blood and skin chimaerism	ABO, MN, S, Rh	1	2	Moores (1966) cited by Race & Sanger (1968)
Blood chimaerism Hermaphroditism	XX/XY, ABO, Duffy, Lutheran, haptoglobin	2	2	Corey *et al.* (1967)
Blood chimaerism Hermaphroditism	XX/XY Haptoglobin	1	2	Park & Jones (1970)
Blood chimaerism Hermaphroditism	XX/XY K	1	2	Eberle *et al.* (1972)
Blood and gonad chimaerism Hermaphroditism	XX/XY, HL-A, Xg^a, MN, S, ADA, chromo- some markers	2	2	De la Chapelle *et al.* (1974)
Blood, skin fibro- blasts, liver, gonad, iris chimaerism Hypospadias	2n (XX)/3n (XXY) Rh, Jk, K	2	2	(e.g.) Lejeune *et al.* (1967)

junctional mitotic events, giving rise to an XX and XY cell line, and the original XXY line could then be lost. In the absence of additional evidence (for example, from blood groups), it is therefore impossible to be sure that XX/XY individuals are chimaeras rather than mosaics. Several cases of this nature have been reported (Overzier, 1964; Bain & Scott, 1965; Brøgger & Aagenaes, 1965; Manuel, Allie & Jackson, 1965; Segni & Grossi-Bianchi, 1965; Lejeune *et al.*, 1966; Deminatti & Maillard, 1967; Fitzgerald, Brehaut,

TABLE 12. *Spontaneous primary chimaeras in other animals*

Species	Evidence of chimaerism	Cell line markers	No. of contributions demonstrated		Reference
			Maternal	Paternal	
Cat	Coat colour Tissue chimaerism	2n (XX)/3n (XXY)	—	—	(e.g.) Chu *et al.* (1964)
Goat	Blood chimaerism Hermaphroditism	XX/XY Chromosome marker	—	—	Padeh *et al.* (1965)
Mouse	Coat colour Germ line	*kr A^y un we*	2	1	Russell & Woodiel (1966)
Mink	Hermaphroditism Tissue chimaerism	2n (XX)/3n (XXY)	—	—	Nes (1966)
Horse	Blood and gonad chimaerism Hermaphroditism	XX/XY A, Fr_3	—	—	Basrur *et al.* (1970)

Shannon & Angus, 1970; Grace, Quantock & Vinik, 1970; Papp, Gardo, Herpay & Arvay, 1970; Barta, Hittner & Regöly-Mérei, 1971; Kakati, Sharma, Udupa & Chaudhuri, 1971; Shanfield, Young & Hume, 1973).

More informative are XX/XY individuals that have two red cell populations carrying different red cell markers, or two different haptoglobin types in the serum. Sometimes the two red cell populations can be separated, using for example differential agglutination for ABO antigens, and the two populations shown to differ at other blood group loci. Such individuals cannot have arisen from a single zygote, so must be considered as chimaeras; if there is evidence (e.g. hermaphroditism) that both cell populations are also present in solid tissues, they may be regarded as primary chimaeras. Table 11 lists the cases reported to date in Man including a few that are not XX/XY. The two cell populations in the individual described by Lejeune *et al.* (1967) proved to be diploid (XX) and triploid (XXY) respectively, and differed at three blood group loci; in the eyes, the iris showed pigment sectoring. In one case (Moores, 1966), both cell populations were XX, and the woman (Mrs T. R., 'a Tamil of good caste') not only showed no evidence of hermaphroditism but was fertile: the presence of two red cell populations was detected during routine ante-natal blood group typing. Her skin was mottled, with lighter and darker patches, a condition seen also in the XX/XY man (Fig. 40) described by Zuelzer *et al.* (1964); presumably the two components also differed with respect to pigment genes.

Similar cases have been reported in animals, though more rarely (see Table 12). The female mouse described by Russell & Woodiel (1966) resembled

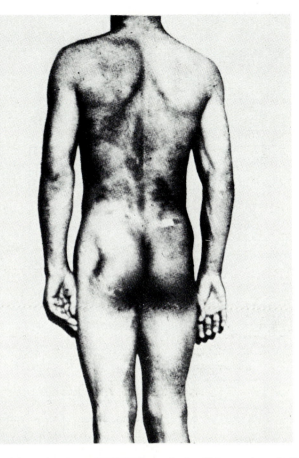

Fig. 40. Skin chimaerism in an XX/XY boy of mixed Negro–Amerindian–Caucasian ancestry. The darker skin, XX in chromosome constitution, grew more slowly in culture than the paler XY skin, and constituted only about 10 % of the total body surface. From Zuelzer *et al.* (1964).

Mrs T.R. in being fertile and having two XX cell lines; it showed yellow and black stripes, since the two lines differed with respect to a coat colour marker, and analysis of the progeny showed that two populations of germ cells were also present, differing at four loci. Male tortoiseshell cats show a bewildering range of sex chromosome abnormalities (for review, see Jones, 1969); some at least are believed to represent primary chimaerism (e.g. Chu, Thuline & Norby, 1964).

It is often possible to show, by comparison of the blood groups of the chimaeric individual and its parents, that either the father or the mother or both have made two genetic contributions to the chimaera, one to each cell

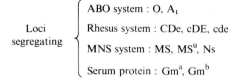

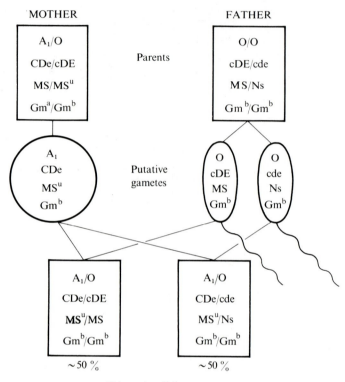

Fig. 41. The inheritance of red cell markers in a chimaeric girl described by Gartler *et al.* (1962), showing evidence of two genetic contributions from the father and one from the mother.

line (see Table 11). The most usual situation is to find two paternal contributions. An example is to be seen in the girl studied by Gartler *et al.* (1962). She had one brown eye and one hazel, and about half her cultured lymphocytes were XX and half XY. She also had two populations of red cells, differentiated at the Rhesus and MNS locus, as shown in Fig. 41. Typing of the parents showed that only the father could have contributed cDE and MS as well as cde and Ns; also, of course, only the father could have contributed

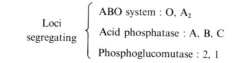

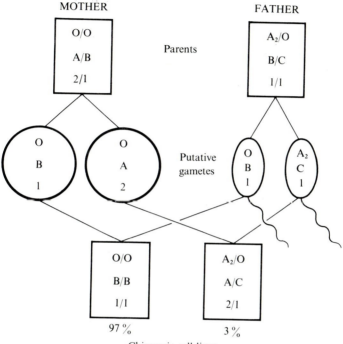

Fig. 42. The inheritance of red cell markers in a chimaeric boy described by Ferguson-Smith *et al.* (1966), showing evidence of two genetic contributions from each parent.

a Y as well as an X chromosome. On the other hand the two chimaeric cell lines were identical for those blood group markers that were segregating in the mother.

Sometimes two maternal as well as two paternal contributions can be demonstrated. Ferguson-Smith *et al.* (1966) studied a 15-year-old true hermaphrodite with bilateral ovotestes and rudimentary uterus and tubes. In culture, 5 % of his lymphocyte mitoses and 73 % of his fibroblast mitoses were XX, the remainder XY. Two classes of red cell, separated by ABO agglutination to give 3 % of group A_2 and 97 % O, were found to be segregating also at the *acid phosphatase* (A, B, C) and *phosphoglucomutase* (1, 2) loci. It turned out (Fig. 42) that only the father could have given A_2 to one cell line and O to the other at the acid phosphatase locus; also only the father could have contributed both an X and a Y chromosome. On the other hand the mother must have given B to one cell line and A to the other at the acid phosphatase locus, and must also have been responsible for the difference seen at the phosphoglucomutase locus.

How might primary chimaerism come about spontaneously? The presence of two separate genetic contributions from the father shows that two spermatozoa must be involved. In theory, these could fertilize (*a*) two separate eggs which then aggregated (type 6); (*b*) the two haploid mitotic products of an exceptional precocious first cleavage division (type 4); (*c*) an egg and second polar body (type 3); (*d*) an egg and first polar body (type 2); (*e*) a secondary oocyte (second meiotic division suppressed) and first polar body (type 1). In all these cases the father, through dispermy, would contribute two independent genomes, while the genetic contributions from the mother could be either independent, identical, or complementary (i.e. products of the same meiotic division). Chimaerism with two maternal and only one paternal contribution could arise if a normal diploid mitotic product of the first cleavage division fused with the second polar body, giving a triploid as well as a diploid cell line (type 5a), or if a secondary oocyte was fertilized (type 5b). These various possibilities, and their genetic consequences, are listed in Table 13. (See Fig. 43 for a diagram of segregation in oogenesis.)

Because chiasma formation occurs at the four-strand rather than the two-strand stage, i.e. between chromatids rather than chromosomes, the likelihood of a given heterozygous locus segregating at the second rather than the first meiotic division depends on its distance from the centromere (Fig. 43). The further a locus is from the centromere, the more likely is segregation to be delayed until the second meiotic division. Thus if a single chiasma is formed midway along the chromosome arm, a secondary oocyte and its first polar body will differ genetically for loci near the centromere, while an ovum and its second polar body will be identical for the same loci; for loci further from the centromere, the reverse will be true. As the number of chiasmata in an arm increases, the distal loci will increasingly tend to segregate independently of the centromere.

If chimaera formation is by embryo aggregation (type 6, Table 13), there will be a 50 % chance of detecting a double genetic contribution at any locus for which the mother is heterozygous. Double fertilization of products of the first meiotic division will make such detection certain, provided that all the cell lines (four for type 1, three for type 2) can be examined. Double fertilization of products of the second meiotic division (type 3) would reduce the probability of detection to below 50 %, with genes nearest the centromere being least likely to show segregation. It follows that the demonstration of two different maternal contributions to the two cell lines of a chimaera is stronger evidence that the chimaerism is of type 1 or 2 rather than type 3 if the locus

TABLE 13. *A classification of primary chimaeras* (*modified from Ford, 1969, table III*)

Type	No. of independent paternal contributions (i.e. no. of sperm)	Mechanism	Genotypes of maternal contributions	Maximum no. of genetically distinct cell lines	Example
1	2	Suppression of second meiotic division; fertilization of secondary oocyte and first polar body	$\dfrac{aB}{ab}\ \dfrac{AB}{Ab}$	4	Not found
2	2	Fertilization of ovum and first polar body	$\begin{matrix}aB\\ \text{or}\\ ab\end{matrix}\ \dfrac{AB}{Ab}$ or $\begin{matrix}AB\\ \text{or}\\ Ab\end{matrix}\ \dfrac{aB}{ab}$	3	Not found
3	2	Fertilization of ovum and second polar body	$\begin{matrix}Ab & AB\\ \text{or}\\ ab & aB\end{matrix}$	2	(e.g.) Zuelzer *et al.* (1964)
4	2	Fertilization of first two mitotic products of unfertilized ovum	One only	2	(e.g.) Gartler *et al.* (1962)
5a	1	Fusion of one mitotic product of fertilized ovum with second polar body	$\begin{matrix}Ab & AB\\ \text{or}\\ ab & aB\end{matrix}$	2	Schmid & Vischer (1967)
5b	1	Suppression of second meiotic division; fertilization of secondary oocyte	One only, e.g. $\dfrac{AB}{Ab}$	2	Not found (but see Donahue, 1972)
6	2	Early aggregation of two embryos	Two, unrelated	2	De la Chapelle *et al.* (1974)

A,a and *B,b* represent alleles at marker loci at which the mother is heterozygous, with *A,a* segregating at the first and *B,b* at the second meiotic division (see Fig. 43).

or loci for which the mother was heterozygous happen to be near the centromere. With type 4, the maternal contribution must always be single.

The weight of evidence in each case can be assessed by considering the number of loci in the mother for which a double maternal contribution could have been demonstrated. These comprise all loci at which the mother is heterozygous, provided the father does not possess the same two alleles, since it would then be impossible to tell from which parent the double contribution came. For example, on this criterion (McLaren, 1969) we find four loci

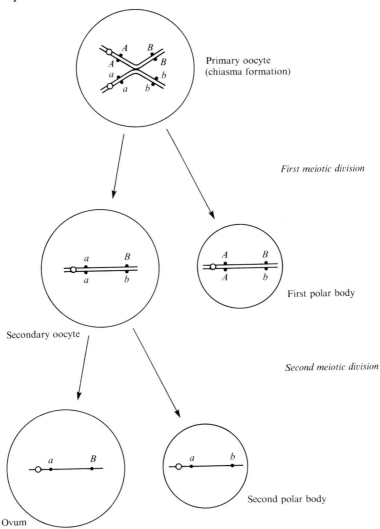

Fig. 43. Segregation of genes during oogenesis, in relation to their distance from the centromere. *A,a* and *B,b* represent alleles at marker loci at which the mother is heterozygous, with *A,a* nearer to and *B,b* further from the centromere (centromeres indicated by circles). The diagram illustrates the situation where one chiasma has formed, between the two loci.

'at risk' in the mother of the chimaeric individual reported by Gartler *et al.* (1962), and five in that of Myhre *et al.* (1965), with two cell lines detected in each and no indication of a double maternal contribution in either. Such a result is fully compatible with type 4 in Table 13, and fits better with type 3 than with 1 or 2. Types 1 and 2 are rendered even more unlikely in the case studied by Myhre *et al.* by the fact that one of the loci demonstrating a single maternal contribution is *Duffy*, which has been shown to be very close indeed (10 centiMorgans (cM)) to the centromere of chromosome 1 (Cook *et al.*, 1974). In both cases, there is also only one maternal contribution for the segregating *Rhesus* locus, which is remote (109 cM) from the centromere of chromosome 1. This favours type 4 rather than 3, unless a double cross-over has taken place, proximal to this locus. Since the phosphoglucomutase locus is also distant from the centromere (79 cM), the double maternal contribution found by Ferguson-Smith *et al.* (1966) favours type 3 rather than 1 or 2.

Perhaps the most plausible mechanism biologically is type 3, the fertilization of an egg by one spermatozoon and the second polar body by another, with incorporation of both sets of cleavage products into the developing embryo. Braden (1957) showed that, in the mouse, the second meiotic division (and even occasionally the first) sometimes results in the formation of two cells of almost equal size ('immediate cleavage'), instead of one large ootid and a smaller polar body. This happens particularly often when fertilization has been delayed and the egg is stale, as the mitotic spindle tends to move into the centre of the egg; delay of fertilization also tends to favour penetration of the egg by two spermatozoa instead of the usual singleton. Perhaps penetration of the egg by more than one spermatozoon may actually precipitate 'immediate cleavage'.

When a double maternal contribution has been demonstrated at several of the loci 'at risk' (e.g. 3/3 in Corey *et al.*, 1967; 2/2 in Ferguson-Smith *et al.*, 1966), type 4 is eliminated, and 1, 2 or 6 become more likely than 3. Fertilization of products of the first meiotic division (types 1 and 2) should give rise to more than two cell populations (see Table 13); such a case has not yet been reported, but of course one or more populations may be lost during embryonic development. Spontaneous embryo aggregation (type 6) has not been demonstrated in any species: we do not know in Man whether the zona pellucida is ever lost early enough to allow aggregation to occur.

The possibility of aggregation of cleaving embryos is sharply raised by the case reported recently by De la Chapelle *et al.* (1974), of an XX/XY female hermaphrodite. Red cell antigens, red cell enzyme types and HL-A antigens showed that two genetically distinct cell populations were present, and that there had been two genetic contributions both from the father and from the mother. Chromosomal markers were also examined, using quinacrine fluorescence banding. The segregation of a maternal marker on chromosome 3 indicated that the two maternal nuclei were probably not the ovum and its second polar body, as the band was very near the centromere and hence crossing-over and segregation at the second meiotic division were very unlikely to have occurred. Mechanisms involving fertilization of products of the first meiotic division were also thought unlikely, since no more than two cell lines were detected. The weight of evidence was thus reckoned to point to aggregation of two independent embryos during cleavage; however, fertilization of products of the first meiotic division, followed by loss of one or more cell lines during development, seems at least as likely.

A similar argument has been used by Russell & Woodiel (1966) in favour of embryo aggregation as the explanation for their spontaneous mouse chimaera; but again, double fertilization of the products of either the first or second meiotic division is a possibility, as also is single fertilization of the secondary oocyte (type 5b), since a double paternal contribution was not shown.

It seems likely that more than one of the mechanisms outlined in Table 13 will need to be invoked to explain the origin of spontaneous chimaeras. Study of a wider range of genetical markers, and better knowledge of their linkage relations, coupled with the identification of the newly discovered centromeric markers, may reveal the probable mechanism in any particular case.

A striking difference between the spontaneous human chimaeras listed in Table 11 and the experimental chimaeras produced in mice is the high incidence of hermaphroditism in human XX/XY chimaeras, and the rarity of this condition in mice (less than 2 %; see Chapter 4). The difference is unlikely to be only one of ascertainment. Of the eleven chimaeras listed in Table 11, eight were subjected to chromosome examination because of ambiguities of their external genitalia or other evidence of sexual abnormality, but in the remaining three the chimaerism was detected in the course of blood group testing. Chromosome examination showed one to be XX/XX (Moores, 1966) and the other two to be XX/XY (Zuelzer *et al.*, 1964; Myhre *et al.*, 1965). One of the two XX/XY individuals, a volunteer blood donor, had been operated on at an early age for an intersexual state of his external genitalia (Zuelzer *et al.*, 1964); the other, blood-grouped because he was undergoing a heart operation, showed hypospadias (Myhre *et al.*, 1965).

Since intersexual features were present in both the XX/XY men who had been identified through blood-grouping, without knowledge of their genital anatomy, it seems unlikely that there exist numerous as yet undetected XX/XY chimaeras among normal fertile men. The contrast with the mouse situation is therefore likely to be real. The spontaneous mouse chimaera described by Russell & Woodiel (1966) unfortunately sheds no light on the problem, as both its cell lines were XX. However, there seems no reason why spontaneous chimaeras in which the two cell lines arise at fertilization should develop differently from experimental chimaeras made at the 8-cell stage; a species difference is therefore likely to be involved. Perhaps in Man the mechanism of sex determination is less rigidly canalized than in the mouse, so that the presence of XX as well as XY cells in the developing gonad induces a state of intersexuality. Little information is available on other species (Table 12).

12

Perspectives

Writing about a rapidly moving field of research is like running up a coming-down escalator. By the time this book is published it will already be out-of-date. Some of what I have written (I hope not too much) will actually have been proved wrong; some will merely have been overtaken by the rapid march of research papers.

Particularly presumptious must be any attempt to answer the question 'Where do we go from here'? The only safe answer, 'Wait and see', is inexcusably dull. Let me then give a very personal answer, by outlining the problems that I would like to see tackled if I had at my disposal a factory for the production of experimental chimacras and an army of research workers to study them. By the time you read this chapter, some of these problems may already have been solved. Others will remain unsolved, perhaps because they are technically too difficult at the present time, perhaps because my enthusiasm for them is not shared by my colleagues.

A quantitative description of mammalian ontogeny must include data on clonal proliferation and cell migration throughout development. In principle, experimental chimaeras provide ideal material for such an analysis, particularly in the early stages of development when the fate of a single donor cell and its progeny can be traced. In practice, however, this will require either the discovery of new cell markers, as specified in Chapter 2, or the laborious application of existing immunological techniques.

Together with a 'fate map' approach to normal development should be linked a study of the ways in which the embryo is able to overcome perturbations of development. An obvious example is the twofold, or more, increase in initial cell number in aggregation chimaeras, posing a problem in growth regulation that the embryo solves within a few days. The possibilities of combining cleaving embryos of different ages, so as to follow the relative contribution of older and younger stages to subsequent development, have yet to be explored. Still more intriguing is the suggestion (Brinster, 1974) that teratocarcinoma cells, or even cells from adult tissues, may become incorporated into the developing embryo following injection into the blastocyst. Would a single differentiated cell have any effect on development? If so, would it retain its initial direction of differentiation, so that a donor lymphocyte for example would take part in the development of the embryo's immune system, or could it switch direction? What about cancer cells?

The elegant use of mosaics in *Drosophila* to pinpoint the primary site of action of mutant genes, especially behavioural mutants (Hotta & Benzer, 1972), depends on the lack of cell migration and the absence of interactions between cells that constitute such a striking feature of insect development. Mammals are different in these respects: nonetheless chimaeras can in principle be used to indicate the main focus of activity of a gene. An example is seen in the study of chimaeras involving muscular dystrophy in mice (Peterson, 1974), which established that the muscle was not in fact the site of action of the gene. Hotta & Benzer's work will probably induce studies of behaviour in mouse chimaeras also.

An unsolved problem of sexual differentiation in mammals is the identity of the actual sex-determining elements within the proto-gonad. If the gonad develops as a testis, the individual (whatever its chromosome constitution) develops as a male: if not, a female phenotype results. The gonad develops as a testis if the chromosome constitution of the cells within it includes a Y chromosome. But which cells? Germ cells, somatic cells, or both? Some of the arguments have been reviewed in Chapter 4; the question is still open. A study of aggregation chimaeras at the earliest embryonic age at which gonadal sex can be determined might throw light on the problem, since the use of appropriate immunological and enzyme markers in XX/XY chimaeras would in principle permit the proportions of XX and XY cells to be determined in germ cells and somatic cells separately. Inclusion of a mutant such as *Steel*, which greatly reduces the number of germ cells reaching the ovary, would sharpen the resolution of the system.

A related question concerns the pattern of migration of germ cells in the genital ridges. In male embryos, at least in the sheep, the germ cells are mostly located in the centre of the gonad at 32 days' gestation; in the female, on the other hand, the germ cells are concentrated in the cortex of the developing gonad. If, in the genital ridges of XX/XY chimaeras, XX and XY germ cells could be distinguished according to their genetic origin, one could observe directly whether their migration pattern was dictated by their own sex chromosome constitution or by that of their cellular environment.

Intersexual conditions are strikingly rare in mouse chimaeras, even in XX/XY individuals. Evidently sexual differentiation is very well canalized, with an effective early switch mechanism. More understanding of the circumstances in which canalization fails, and XX/XY chimaeras develop as hermaphrodites, might throw light on the causes of intersexuality in Man.

Cell lineage studies could also throw light on basic immunological issues. Lymphoid stem cells differentiate into T and B cells; they also differentiate into populations of cells producing antibody to different antigens. Which differentiation occurs first? In a chimaera, correlations could be looked for in the relative proportions of the two components (identified by allotype markers) in the T and B cell populations of various antigenic specificities.

Thus, if a C3H↔C57BL chimaera showed around 80 % of allotype H-2^b (C57BL) in both T and B cells reacting with one particular antigen, with other percentages for other antigens, it would suggest that antigenic specificities had become established in lymphoid ontogeny before the divergence of T and B cell lines. Conversely, if all T cell populations resembled one another more closely than they did the antigenically corresponding B cell types, one could conclude that the allocation of T and B cells had occurred before the appearance of antigenic specificity.

The phenomenon of allelic exclusion in antibody-forming cells may give rise to mosaicism in heterozygous animals. A comparison of such mosaics with appropriate chimaeras might yield information on the relationship between allelic exclusion and the establishment of immunological specificity.

As mentioned in Chapter 7, there is at present no unanimity as to the basis for mutual unresponsiveness between the component lymphomyeloid populations in chimaeras. The concept of primary tolerance has been challenged, but the demonstration of blocking antibody has yet to be experimentally confirmed. The present conflict of evidence is particularly unsatisfactory, since the mechanisms underlying such a fundamental phenomenon as self-recognition seem unlikely to be strain-specific or even species-specific. If, as has been suggested, some strain combinations are genetically incompetent at developing or maintaining mutual tolerance, analysis of the causes of this defect could throw light on the aetiology of autoimmune diseases in Man.

When we turn to phenotypic interactions between genetically distinct cell populations, the value of chimaeras becomes still more obvious. Measurements of patch size, as have been made on the pigmented retina for example, can now be converted to estimates of clone size; these provide a sensitive indicator of the tendency of cells to stick together. One chimaeric strain combination has so far been identified in which cells of like genotype tend preferentially to stick together. This is particularly striking, since tests of cell aggregation *in vitro* have in the main shown differential aggregation with respect to tissue origin (e.g. neural retina, heart, kidney) rather than genotype. Presumably differential aggregation reflects differences in cell surface conformation; since surface similarities may be required for normal cell interaction, one wonders whether chimaeras of some genetically divergent strain combinations may turn out to be not entirely representative of normal embryonic development.

For experimental embryology studies, using genetic differences between chimaera components as markers to facilitate the analysis of cell lineage, the residual genetic differences between components should be as small as possible so as to ensure that development is as normal as possible. Congenic strains, differing only at a single locus, provide ideal material. When we are considering the role of genetic differences in cell-to-cell interaction, on the other hand, genetic divergence between the components may prove more informative than genetic similarity. The extreme of genetic divergence is provided by interspecific chimaeras. This is a field in which very little has so far been done, but which shows immense promise. *In vitro*, cells from species as remote as chick and mouse have been shown to co-operate morphogenetically, for example in the reconstruction of kidney tubules from dissociated cells.

For interspecific studies, injection chimaeras are more suitable than aggregation chimaeras because the chimaerism is then limited to the embryo and is less likely to affect the interaction between trophoblast and uterus. The causes of interspecific hybrid breakdown (e.g. sheep × goat, *Peromyscus maniculatus × P. polionotus*) might be illuminated by combining the techniques of interspecies egg transfer and the injection of whole inner cell masses into trophoblast vesicles: the three variables of mother, trophoblast and embryo could in this way be varied independently.

Sex chromosomes affect not only the direction of differentiation of the gonad and hence the individual, but also the development of the germ cells. Some interaction appears to take place between the sex chromosome constitution of the germ cells and the cellular environment in which they find themselves. For example, chimaera studies indicate that XX germ cells in a testis, in contrast to XY germ cells, enter meiosis before birth and fail to undergo spermatogenesis; but other lines of evidence have suggested that XO germ cells in a testis may be much more successful, even forming mature spermatozoa. Perhaps the requirement for normal germ cell development in the testis is absence of a second X chromosome rather than presence of a Y chromosome. Chimaeras made between XO and XY embryos should be able to confirm this hypothesis. There is also the intriguing case of the XY egg shed from the XX↔XY chimaeric ovary: did an initially XY germ cell succeed in undergoing oogenesis, only to suffer non-disjunction at the first meiotic division, or was the non-disjunctional event earlier, giving an XXY germ cell that survived in the ovary by virtue of its second X chromosome? If the second explanation is true, how do XO germ cells manage to differentiate into normal eggs? There is some evidence to indicate that the eggs developed from XO germ cells are not in fact normal (P. S. Burgoyne, personal communication). The role of X and Y chromosomes in the development of germ cells as well as in sexual differentiation is an obvious target for chimaera studies.

The factors determining susceptibility and resistance to tumours are of more than academic interest. In chimaeras between susceptible and resistant strains, the development of tumours seems often to be cell-autonomous, but by no means always. In chimaeras of at least one strain combination, lymphoma susceptibility is partially suppressed, despite the fact that most of the tissues contain an overwhelming majority of cells of the susceptible strain. How can so few resistant cells control such a large susceptible population? Perhaps some diffusible resistance-inducing substance exists. Again, why is it that the tumours developing in chimaeras between susceptible and resistant strains come sometimes from cells of the resistant component? Obviously some environmental influence must be at work: is it the biochemical or hormonal environment that is involved, or an oncogenic virus, or perhaps some contact-mediated change in cell surface properties?

An area of genetic interaction that remains entirely unexplored is the possibility of 'vegetative heterosis'. When sexual crosses are made between different strains of animals, the progeny often show 'hybrid vigour' or heterosis with respect to such characters as growth rate, milk production and reproductive performance. Although the physiological basis for heterosis is not yet fully understood, one factor is thought to be the extra biochemical flexibility made possible by heterozygosity. Each allelic version of an enzyme may be advantageous in different physiological circumstances; when two alternatives are available the range of situations with which the cell can cope is enhanced. By analogy the presence of two genetically different populations of cells in one individual might also confer biochemical flexibility. All those who have raised experimental chimaeras bear witness to their vigour, but no quantitative assessment of chimaeras relative to their component strains has been made with respect to characters showing heterosis in sexual crosses.

Inherent in the concept of vegetative heterosis at the cellular level is the idea of 'metabolic co-operation', in which a metabolic deficiency in one cell is corrected by gene products emanating from neighbouring cells. Such co-operative functions have been demonstrated in cultures of cells of assorted genotype (Subak-Sharpe, Bürk & Pitts, 1969). Low molecular weight substances may diffuse from cell to cell; larger molecules may pass through specialized areas of the cell membrane ('gap junctions') formed when processes from neighbouring cells make contact with one another. No information is available as to whether in chimaeric tissues gap junctions form equally between cells of like and unlike genotype.

The value of chimaeras for analytical studies is great. Will they ever prove of any direct practical use, for example in agriculture? Embryos homozygous for genes that are normally lethal can sometimes be 'rescued' by aggregating them with normal embryos: this procedure not only distinguishes between autonomous cell-lethals (not amenable to rescue) and lethals acting at the level of the tissue or organism, but can be used for linkage testing of lethals impossible by any other means. This technique has already been used to establish the allelism of two X-linked genes, *jimpy* and *myelin synthesis deficiency*. The injection of quite a small number of cells into an early embryo might cure genetic constitutions that would otherwise be lethal or sterile.

On a more speculative level, there exists a further potential application that might arise in the remote future. If vegetative propagation, for example by nuclear transfer, became a routine procedure in any animal species, it might then become desirable to build up highly selected gene combinations for advantageous traits, involving heterozygosity at numerous loci, as has been done in plant breeding. A single meiotic event, with its resultant multiple recombination, would destroy many years' work in building up the optimal genome. Yet there might be circumstances in which it was desired to combine two such genomes in the same individual, perhaps to exploit vegetative heterosis if such a phenomenon exists, perhaps just to see how two sets of characters would combine together as a guide to whether it would be worth bringing them together genetically. Chimaerism would provide the obvious means, allowing genomes to be combined without the necessity for meiosis. An analogous situation already exists in plants, where highly selected plant varieties are often propagated by vegetative means alone, to avoid meiosis and recombination, and where 'graft hybridization' is a standard procedure for combining valuable traits within a single individual.

Such procedures applied to animals are still, however, in the realms of science fiction. The importance of chimaeras in science today lies in their use as an analytical tool in experimental embryology and developmental genetics. I have tried in this book to give an account of what has been done so far: I hope I have also given some impression of how much is still waiting to be done.

Note added in proof

For use of chimaeras to 'rescue' a sterile genotype (*Tfm*/Y), see:

> Lyon, M.F., Glenister, P.H. & Lamoreux, M.L. (1975). Normal spermatozoa from androgen-resistant germ cells of chimaeric mice and the role of androgen in spermatogenesis. *Nature, Lond.* **258**, 620–2.

For the integration of tumour cells (i.e. teratocarcinoma) into normal embryonic development by the injection chimaera technique, see:

> Papaioannou, V.E., McBurney, M.W., Gardner, R.L. & Evans, M.J. (1975). Fate of teratocarcinoma cells injected into early mouse embryos. *Nature, Lond.* **258**, 70–2.

> Mintz, B. & Illmensee, K. (1975). Normal genetically mosaic mice produced from malignant teratocarcinoma cells. *Proc. Nat. Acad. Sci., USA* **72**, 3585–9.

Bibliography

Allison, A. C. (1971). Unresponsiveness to self antigens. *Lancet*, ii, 1401–3.

Ansell, J. D., Barlow, P. W. & McLaren, A. (1974). Binucleate and polyploid cells in the decidua of the mouse. *J. Embryol. exp. Morph.* **31**, 223–7.

Bain, A. D. & Scott, J. S. (1965). Mixed gonadal dysgenesis with XX/XY mosaicism. *Lancet*, i, 1035–9.

Baker, W. W. & Mintz, B. (1969). Subunit structure and gene control of mouse NADP-malate dehydrogenase. *Biochem. Genet.* **2**, 351–60.

Barnes, R. D. (1974). Leukaemia – an immune deficiency? *Adv. Biosci.* **12**, 160–8.

Barnes, R. D., Holliday, J. & Tuffrey, M. (1974a). Immunofluorescence and elution studies in tetraparental NZB↔CFW chimaeras and graft-versus-host diseased NZB. *Immunology* **26**, 1195–206.

Barnes, R. D. & Tuffrey, M. (1972a). Tolerance in ovum fusion derived tetraparental mouse chimaeras. In *Advances in Experimental Medicine and Biology*, ed. B. D. Jankovic, vol. 29, pp. 427–34. New York & London: Plenum Press.

Barnes, R. D. & Tuffrey, M. (1972b). Immunity in ovum fusion derived tetraparental mice. *Scand. J. Immunol.* **1**, 284. (Abstr.)

Barnes, R. D. & Tuffrey, M. (1973). Allotype-specific anti-autoantibody activity in tetraparental NZB mice. *Eur. J. Immunol.* **3**, 60–1.

Barnes, R. D. & Tuffrey, M. (1974). Lymphoma susceptibility of the AKR acquired prior to the stage of implantation. *Brit. J. Cancer* **29**, 400–2.

Barnes, R. D., Tuffrey, M., Drury, L. & Catty, D. (1974b). Unequal rates of cell proliferation in tetraparental mouse chimaeras derived by fusion of early embryos. *Differentiation* **2**, 257–60.

Barnes, R. D., Tuffrey, M. & Ford, C. E. (1973). Suppression of lymphoma development in tetraparental AKR mouse chimaeras derived from ovum fusion. *Nature New Biol.* **244**, 282–3.

Barnes, R. D., Tuffrey, M., Graham, C. F., Holliday, J. & Thornton, C. (1974c). Reduced levels of serum haemolytic complement and renal lesions in ovum-fusion-derived tetraparental mouse chimaeras. *Scand. J. Immunol.* **3**, 789–96.

Barnes, R. D., Tuffrey, M. & Kingman, J. (1972b). The delay of leukaemia in tetraparental ovum fusion-derived AKR chimaeras. *Clin. exp. Immunol.* **12**, 541–5.

Barnes, R. D., Tuffrey, M., Kingman, J., Thornton, C. & Turner, M. W. (1972a). The disease of the NZB mouse. I. Examination of ovum fusion derived tetraparental NZB:CFW chimaeras. *Clin. exp. Immunol.* **11**, 605–28.

Barnes, R. D., Wills, E. J. & Tuffrey, M. (1975). The disease of the NZB mouse. II. Further studies on tetraparental NZB↔CFW chimaeras derived by aggregation of early embryos. *J. Immunogenetics* **2**, 195–209.

Barta, von L., Hittner, I. & Regöly-Mérei, F. A. (1971). Pseudohermaphroditismus masculinus mit XX/XY Mosaizismus. *Arch. Kinderheilk.* **183**, 353–9.

Basrur, P. K., Kanagawa, H. & Podliachouk, L. (1970). Further studies on the cell populations of an intersex horse. *Can. J. comp. Med.* **34**, 294–8.

Beatty, R. A. (1972). The genetics of size and shape of spermatozoan organelles. In *The Genetics of the Spermatozoon*, ed. R. A. Beatty & S. Gluecksohn-Waelsch, pp. 97–115. Edinburgh & New York.

Bechtol, K. B., Freed, J. H., Herzenberg, L. A. & McDevitt, H. O. (1974). Genetic control of the antibody response to poly-L (Tyr, Glu)-poly-D,L-Ala - - poly-L, Lys in C3H↔CWB tetraparental mice. *J. exp. Med.* **140**, 1660–75.

Bechtol, K. B., Wegmann, T. G., Chesebro, B. W., Herzenberg, L. A. & McDevitt, H. O. (1971). The antibody response to (T,G)-A-L in tetraparental mice. *Fed Proc.* **30**, 1531. (Abstr.)

Bechtol, K. B., Wegmann, T. G., Freed, J. H., Grumet, F. C., Chesebro, B. W., Herzenberg, L. A. & McDevitt, H. O. (1974). Genetic control of the immune response to (T,G)-A--L in C3H↔C57 tetraparental mice. *Cell. Immunol.* **13**, 264–77.

Benirschke, K. (1970). Spontaneous chimerism in mammals: a critical review. *Current Topics in Pathology* **51**, 1–61.

Biggers, J. D., Whitten, W. K. & Whittingham, D. G. (1971). The culture of mouse embryos *in vitro*. In *Methods in Mammalian Embryology*, ed. J. C. Daniel, pp. 86–116. San Francisco: Freeman.

Blackler, A. W. (1962). Transfer of primordial germ-cells between two subspecies of *Xenopus laevis*. *J. Embryol. exp. Morph.* **10**, 641–51.

Blackler, A. W. & Fischberg, M. (1961). Transfer of primordial germ-cells in *Xenopus laevis*. *J. Embryol. exp. Morph.* **9**, 634–41.

Bona, C. A., Tuffrey, M. & Barnes, R. D. (1974). Distribution of theta antigens in ovum fusion derived AKR↔CBA/H-T6T6 tetraparental mouse chimaeras. *Tissue Antigens*, **4**, 31–5.

Booth, P. B., Plaut, G., James, J. D., Ikin, E. W., Moores, P., Sanger, R. & Race, R. R. (1957). Blood chimerism in a pair of twins. *Brit. Med. J.* i, 1456–8.

Bowman, P. & McLaren, A. (1970). Viability and growth of mouse embryos after *in vitro* culture and fusion. *J. Embryol. exp. Morph.* **23**, 693–704.

Braden, A. W. H. (1957). Variation between strains in the incidence of various abnormalities of egg maturation and fertilization in the mouse. *J. Genet.* **55**, 476–86.

Brinster, R. L. (1963). A method for *in vitro* cultivation of mouse ova from two-cell to blastocyst. *Exp. Cell Res.* **32**, 205–8.

Brinster, R. L. (1965a). Studies on the development of mouse embryos *in vitro*. II. The effect of energy source. *J. exp. Zool.* **158**, 59–68.

Brinster, R. L. (1965b). Studies on the development of mouse embryos *in vitro*. III. The effect of fixed nitrogen source. *J. exp. Zool.* **158**, 69–77.

Brinster, R. L. (1974). The effect of cells transferred into the mouse blastocyst on subsequent development. *J. exp. Med.* **140**, 1049–56.

Brøgger, A. & Aagenaes, O. (1965). The human Y chromosome and the etiology of true hermaphroditism. *Hereditas* **53**, 231–46.

Brøgger, A. & Gudersen, S. K. (1966). Double fertilization in Down's syndrome. *Lancet* i, 1270–1.

Buehr, M. & McLaren, A. (1974). Size regulation in chimaeric mouse embryos. *J. Embryol. exp. Morph.* **31**, 229–34.

Burgoyne, P. S. (1973). The genetic control of germ cell differentiation in mice. PhD thesis, University of Edinburgh.

Burgoyne, P. S. (1975). Sperm phenotype and its relationship to somatic and germ line genotype. A study using mouse aggregation chimaeras. *Dev. Biol.* **44**, 63–76.

Burnet, F. M. (1959). *The Clonal Selection Theory of Acquired Immunity.* London: Cambridge University Press.

Cattanach, B. M. (1974). Position effect variegation in the mouse. *Genet. Res.* **23**, 291–306.

Cattanach, B. M., Pollard, C. E. & Hawkes, S. G. (1971). Sex reversed mice: XX and XO males. *Cytogenetics* **10**, 318–37.

Cattanach, B. M., Wolfe, H. G. & Lyon, M. F. (1972). A comparative study of the coats of chimaeric mice and those of heterozygotes for X-linked genes. *Genet. Res.* **19**, 213–28.

Ceppellini, R. (1971). Old and new facts and speculations about transplantation antigens of Man. In *Progress in Immunology*, ed. B. Amos, pp. 973–1025, New York & London: Academic Press.

Chang, M. C. (1950). Development and fate of transferred rabbit ova or blastocysts in relation to the ovulation time of recipients. *J. exp. Zool.* **114**, 197–216.

Chapman, V. M., Ansell, J. D. & McLaren, A. (1972). Trophoblast giant cell differentiation in the mouse: expression of glucose phosphate isomerase (GPI-1) electrophoretic variants in transferred and chimaeric embryos. *Dev. Biol.* **29**, 48–54.

Chu, E. H. Y., Thuline, H. C. & Norby, D. E. (1964). Triploid–diploid chimerism in a male tortoiseshell cat. *Cytogenetics* **3**, 1–18.

Cleffman, G. (1963). Agouti pigment cells *in situ* and *in vitro. Ann. N.Y. Acad. Sci.* **100**, 749–61.

Cohn, M. (1972). Discussion. In *Genetic Control of Immune Responsiveness*, ed. H. O. McDevitt & M. Landy, pp. 441–444. New York & London: Academic Press.

Condamine, H., Custer, R. P. & Mintz, B. (1971). Pure-strain and genetically mosaic liver tumors histochemically identified with the β-glucuronidase marker in allophenic mice. *Proc. Nat. Acad. Sci. USA,* **68**, 2032–6.

Cook, P. J. L., Robson, E. B., Buckton, K. E., Jacobs, P. A. & Polani, P. E. (1974). Segregation of genetic markers in families with chromosome polymorphisms and structural rearrangements involving chromosome 1. *Ann. Hum. Genet., Lond.* **37**, 261–273.

Corey, M. J., Miller, J. R., Maclean, J. R. & Chown, B. (1967). A case of XX/XY mosaicism. *Am. J. Hum. Genet.* **19**, 378–87.

Coulombre, J. C. & Russell, E. S. (1954). Analysis of the pleiotropism at the *W*-locus in the mouse. The effects of *W* and *W^v* substitution upon postnatal development of germ cells. *J. exp. Zool.* **126**, 277–95.

Curtis, A. S. G. (1967). *The Cell Surface – its Molecular Role in Morphogenesis*, p. 251. London: Logos Press.

Dalcq, A. M. (1957). *Introduction to General Embryology.* London: Oxford University Press.

Daniel, J. C. (ed.) (1971). *Methods in Mammalian Embryology*. San Francisco: Freeman.

Daniel, J. C. & Olsen, J. D. (1966). Cell movement, proliferation and death in the formation of the embryonic axis of the rabbit. *Anat. Rec.* **156**, 123–8.

De Grouchy, J., Moullec, C., Salmon, N., Josso, N., Frézal, J. & Lamy, M. (1964). Hermaphrodisme avec caryotype XX/XY. Etude génétique d'un cas. *Ann. Génét.* **7**, 24–30.

De la Chapelle, A., Schröder, J., Rantanen, P., Thomasson, B., Niemi, M., Tiilikainen, A., Sanger, R. & Robson, E. B. (1974). Early fusion of two human embryos? *Ann. Hum. Genet., Lond.* **38**, 63–75.

Deminatti, M. & Maillard, E. (1967). Etude d'un cas d'hermaphrodisme humain à caryotype 46,XY/46,XX. *C. r. Acad. Sci., Paris* **265**, 365–8.

Deol, M. S. (1973). Chimaeras and the forbidden-clone theory of self-tolerance. *Nature, Lond.*, **242**, 469.

Deol. M. S. & Whitten, W. K. (1972*a*). Time of X chromosome inactivation in retinal melanocytes of the mouse. *Nature New Biol.* **238**, 159–60.

Deol, M. S. & Whitten, W. K. (1972*b*). X-chromosome inactivation: does it occur at the same time in all cells of the embryo? *Nature New Biol.* **240**, 277–9.

Dickmann, Z. (1971). Egg transfer. In *Methods in Mammalian Embryology*, ed. J. C. Daniel, pp. 133–45. San Francisco: Freeman.

Donahue, R. P. (1972). Cytogenetic analysis of the first cleavage division in mouse embryos. *Proc. Nat. Acad. Sci. USA* **69**, 74–7.

Drews, U. & Alonso-Lozano, V. (1974). X-inactivation pattern in the epididymis of sex-reversed mice heterozygous for testicular feminization. *J. Embryol. exp. Morph.* **32**, 217–25.

Dunn, G. R. (1972). Expression of a sex-linked gene in standard and fusion-chimeric mice. *J. exp. Zool.* **181**, 1–16.

Eberle, V. P., Gallasch, E. & Truss, F. (1972). XX/XY-hermaphroditismus verus mit Blutchimäre. *Blut*, **25**, 255–64.

Eicher, E. M. & Hoppe, P. C. (1973). Use of chimeras to transmit lethal genes in the mouse and to demonstrate allelism of the two X-linked male lethal genes *jp* and *msd*. *J. exp. Zool.* **183**, 181–4.

Ferguson-Smith, M. A., Izatt, M., Renwick, J. H. & Mack, W. S. (1966). Unpublished; cited in McLaren (1969).

Fitzgerald, P. H., Brehaut, L. A., Shannon, F. T. & Angus, H. B. (1970). Evidence of XX/XY sex chromosome mosaicism in a child with true hermaphroditism. *J. med. Genet.* **7**, 383–8.

Ford, C. E. (1969). Mosaics and chimaeras. *Brit. Med. Bull.* **25**, 104–9.

Ford, C. E. (1970). The cytogenetics of the male germ cells and the testis in mammals. In *The Human Testis*, ed. E. Rosenberg & C. A. Paulsen, pp. 139–49. New York & London: Plenum Press.

Ford, C. E., Evans, E. P., Burtenshaw, M. D., Clegg, H., Barnes, R. D. & Tuffrey, M. (1974). Marker chromosome analysis of tetraparental AKR↔CBA-T6 mouse chimaeras. *Differentiation* **2**, 321–33.

Ford, C. E., Evans, E. P., Burtenshaw, M. D., Clegg, H. M., Tuffrey, M. & Barnes, R. D. (1975*a*). A functional 'sex-reversed' oocyte in the mouse. *Proc. Roy. Soc. Lond. B*, **190**, 187–97.

Ford, C. E., Evans, E. P. & Gardner, R. L. (1975*b*). Marker chromosome analysis of two mouse chimaeras. *J. Embryol. exp. Morph.* **33**, 447–57.

Ford, C. E., Micklem, H. S. & Ogden, D. A. (1968). Evidence for the existence of a lymphoid stem cell. *Lancet* i, 621–2.

Freed, J. H., Bechtol, K. B., Herzenberg, L. A., Herzenberg, L. A. & McDevitt, H. O. (1973). Analysis of anti-(T,G)-A--L antibody in tetraparental mice. *Transplant. Proc.* **5**, 167–71.

Gandini, E., Gartler, S. M., Angioni, G., Argiolas, N. & Dell'Accua, G. (1968). Developmental implications of multiple tissue studies in glucose-6-phosphate dehydrogenase-deficient heterozygotes. *Proc. Nat. Acad. Sci. USA*, **61**, 945–8.

Garcia-Bellido, A., Ripoll, P. & Morata, G. (1973). Developmental compartmentalisation of the wing disk of *Drosophila*. *Nature New Biol.* **245**, 251–3.

Gardner, R. L. (1968). Mouse chimaeras obtained by the injection of cells into the blastocyst. *Nature, Lond.* **220**, 596–7.

Gardner, R. L. (1971). Manipulations on the blastocyst. *Adv. Biosci.* **6**, 279–96.

Gardner, R. L. (1974). Microsurgical approaches to the study of early mammalian development. In *Birth Defects and Fetal Development, Endocrine and Metabolic Factors*, ed. K. S. Moghissi, pp. 212–33. Illinois: C. C. Thomas.

Gardner, R. L. (1975*a*). Analysis of determination and differentiation in the early mammalian embryo using intra- and interspecific chimeras. In *The Developmental Biology of Reproduction, Symposium of the Society for Developmental Biology* 33, pp. 207–36. New York & London: Academic Press.

Gardner, R. L. (1975*b*). Origins and properties of trophoblast. In *Immunobiology of Trophoblast*, ed. R. G. Edwards, C. W. S. Howe & M. H. Johnson, pp. 43–66. London: Cambridge University Press.

Gardner, R. L. & Johnson, M. H. (1972). An investigation of inner cell mass and trophoblast tissues following their isolation from the mouse blastocyst. *J. Embryol. exp. Morph.* **28**, 279–312.

Gardner, R. L. & Johnson, M. H. (1973). Investigation of early mammalian development using interspecific chimaeras between rat and mouse. *Nature New Biol.* **246**, 86–9.

Gardner, R. L. & Johnson, M. H. (1975). Investigation of cellular interaction and deployment in the early mammalian embryo using inter-specific chimaeras between rat and mouse. In *Cell Patterning, Ciba Foundation Symposium* 29, pp. 183–200. Amsterdam: Excerpta Medica, North Holland.

Gardner, R. L. & Lyon, M. F. (1971). X chromosome inactivation studied by injection of a single cell into the mouse blastocyst. *Nature, Lond.* **231**, 385–6.

Gardner, R. L. & Munro, A. J. (1974). Successful construction of chimaeric rabbit. *Nature, Lond.* **250**, 146–7.

Gardner, R. L., Papaioannou, V. E. & Barton, S. C. (1973). Origin of the ectoplacental cone and secondary giant cells in mouse blastocysts reconstituted from isolated trophoblast and inner cell mass. *J. Embryol. exp. Morph.* **30**, 561–72.

Garner, W. & McLaren, A. (1974). Cell distribution in chimaeric mouse embryos before implantation. *J. Embryol. exp. Morph.* **32**, 495–503.

Gartler, S. M., Waxman, S. H. & Giblett, E. R. (1962). An XX/XY human hermaphrodite resulting from double fertilization. *Proc. Nat. Acad. Sci. USA*, **48**, 332–5.

Gearhart, J. D. & Mintz, B. (1971). Single-somite and single-myotome multiclonal development in the mouse. *Amer. Zool.* **11**, 677–8.

Gearhart, J. D. & Mintz, B. (1972a). Clonal origins of somites and their muscle derivatives: evidence from allophenic mice. *Dev. Biol.*, **29**, 27–37.

Gearhart, J. D. & Mintz, B. (1972b). Glucosephosphate isomerase subunit-reassociation tests for maternal–fetal and fetal–fetal cell fusion in the mouse placenta. *Dev. Biol.* **29**, 55–64.

George, J. C. (1958). Experimental fusion of embryos of *Limnaea stagnalis* L. *Koninkl. Nederl. Akademie van Wetenschappen*, Amsterdam Proc., Ser. C **61**, 595–7.

Gornish, M., Webster, M. P. & Wegmann, T. G. (1972). Chimaerism in the immune system of tetraparental mice. *Nature New Biol.* **237**, 249.

Grace, H. J., Quantock, O. P. & Vinik, A. (1970). An unusual cause of 'haematuria' in an XX/XY hermaphrodite. *S. Afr. Med. J.* **44**, 40–3.

Green, M. C. (1968). Mechanism of the pleiotropic effects of the short-ear mutant gene in the mouse. *J. exp. Zool.* **167**, 129–50.

Grüneberg, H. (1966). The molars of the tabby mouse, and a test of the single-active X-chromosome hypothesis. *J. Embryol. exp. Morph.* **15**, 223–44.

Grüneberg, H. (1967). Gene action in the mammalian X-chromosome. *Genet. Res.*, **9**, 343–57.

Grüneberg, H., Cattanach, B. M., McLaren, A., Wolfe, H. G. & Bowman, P. (1972). The molars of tabby chimaeras in the mouse. *Proc. Roy. Soc. Lond.* B., **182**, 183–92.

Grüneberg, H. & McLaren, A. (1972). The skeletal phenotype of some mouse chimaeras. *Proc. Roy. Soc. Lond* B., **182**, 9–23.

Hadorn, E. (1945). Zur Pleiotropie der Genwirkung. *Arch. Jul. Klaus Stift.* **20**, 82–95.

Hadorn, E. (1961). *Developmental Genetics and Lethal Factors*. London: Methuen.

Hammond, J., Jr (1949). Recovery and culture of tubal mouse ova. *Nature, Lond.*, **163**, 28–9.

Heape, W. (1890). Preliminary note on the transplantation and growth of mammalian ova within a uterine foster-mother. *Proc. Roy Soc. Lond.* **48**, 457–8.

Hellström, I., Hellström, K. E. & Allison, A. C. (1971). Neonatally induced allograft tolerance may be mediated by serum-borne factors. *Nature, Lond.*, **230**, 49–50.

Hellström, I., Hellström, K. E., Storb, R. & Thomas, E. D. (1970). Colony inhibition of fibroblasts from chimeric dogs mediated by the dogs' own lymphocytes and specifically abrogated by their serum. *Proc. Nat. Acad. Sci. USA* **66**, 65–71.

Hellström, K. E. & Hellström, I. (1970). Immunological enhancement as studied by cell culture techniques. *Ann. Rev. Microbiol.*, **24**, 373–98.

Hillman, N., Sherman, M. I. & Graham, C. (1972). The effect of spatial arrangements on cell determination during mouse development. *J. Embryol. exp. Morph.*, **28**, 263–78.

Hotta, Y. & Benzer, S. (1972). Mapping of behaviour in *Drosophila* mosaics. *Nature, Lond.* **240**, 527–35.

Johnson, M. H. & Gardner, R. L. (1975). Analysis of rat:mouse chimaeras by immunofluorescence: a preliminary report. In *Immunology in Obstetrics & Gynaecology, Proceedings of 1st International Congress of the Immunology of*

Obstetrics and Gynaecology, Padua, 1973, Ed. A. Centaro & N. Carretti. pp. 312–14. Amsterdam: Excerpta Medica/North Holland.

Jones, T. C. (1969). Anomalies of sex chromosomes in tortoiseshell male cats. In *Comparative Mammalian Cytogenetics*, ed. K. Benirschke, pp. 414–33. Berlin-Heidelberg-New York: Springer Verlag.

Kakati, S., Sharma, T., Udupa, K. N. & Chaudhuri, S. P. R. (1971). A true hermaphrodite with XX/XY mosaicism. *Indian J. Med. Res.* **59**, 104–6.

Kaliss, N. (1969). Immunological enhancement. *Intern. Rev. Exp. Path.*, **8**, 241–76.

Kaliss, N., Whitten, W. K., Wegmann, T. G. & Carter, S. (1974). *Chimeric mice: examples of tolerance or enhancement?* 45th Annual Report, The Jackson Laboratory, Bar Harbor, USA.

Kaufman, M. H. (1973). Analysis of the first cleavage division to determine the sex-ratio and incidence of chromosome anomalies at conception in the mouse. *J. Reprod. Fert.* **35**, 67–72.

Kelly, S. J. (1975). Potency of early cleavage blastomeres of the mouse. In *The Early Development of Mammals, British Society for Developmental Biology* **2**, ed. M. Balls & A. E. Wild, pp. 97–105. London: Cambridge University Press.

Kofman-Alfaro, S. & Chandley, A. C. (1970). Meiosis in the male mouse. An autoradiographic investigation. *Chromosoma* **31**, 404–20.

Lejeune, J., Berger, R., Réthoré, M. O., Vialatte, J. & Salmon, C. (1966). Sur un cas d'hermaphrodisme XX/XY. *Ann. Génét.* **9**, 171–3.

Lejeune, J., Salmon, C., Berger, R., Réthoré, M. O., Rossier, A. & Job, J. C. (1967). Chimère 46, XX/69, XXY. *Ann. Génét.* **10**, 188–92.

Lewis, J. (1973). The theory of clonal mixing during growth. *J. theor. Biol.* **39**, 47–54.

Lewis, J. H., Summerbell, D. & Wolpert, L. (1972). Chimaeras and cell lineage in development. *Nature, Lond.* **239**, 276–8.

Lyon, M. F. (1970). Genetic activity of sex chromosomes in somatic cells of mammals. *Phil. Trans. Roy. Soc. Lond. B*, **259**, 41–52.

Lyon, M. F. (1972). X-chromosome inactivation and developmental patterns in mammals. *Biol. Rev.* **47**, 1–35.

Manuel, M. A., Allie, A. & Jackson, W. P. U. (1965). A true hermaphrodite with XX/XY chromosome mosaicism. *S. Afr. Med. J.* **39**, 411–14.

Marcum, J. B. (1974). The freemartin syndrome. *Anim. Breeding Abstr.* **42**, 227–42.

Marsk, L. & Larsson, K. S. (1974). A simple method for non-surgical blastocyst transfer in mice. *J. Reprod. Fert.* **37**, 393–8.

Marzullo, G. (1970). Production of chick chimaeras. *Nature, Lond.*, **225**, 72–3.

Mayer, J. F., Jr & Fritz, H. I. (1974). The culture of preimplantation rat embryos and the production of allophenic rats. *J. Reprod. Fert.* **39**, 1–9.

Mayer, T. C. & Fishbane, J. L. (1972). Mesoderm–ectoderm interaction in the production of the agouti pigmentation pattern in mice. *Genetics* **71**, 297–303.

McDevitt, H. O. (1972). Discussion. In *Genetic Control of Immune Responsiveness*, ed. H. O. McDevitt & M. Landy, pp. 67–9. New York & London: Academic Press.

McDevitt, H. O., Bechtol, K. B., Grumet, F. C., Mitchell, G. F. & Wegmann, T. G. (1971). Genetic control of the immune response to branched synthetic polypeptide antigens in inbred mice. In *Progress in Immunology*, ed. B. Amos, pp. 495–508. New York & London: Academic Press.

McLaren, A. (1969). Recent studies on developmental regulation in vertebrates. In *Handbook of Molecular Cytology*, ed. A. Lima-De-Faria, pp. 639–655. Amsterdam & London: North-Holland.

McLaren, A. (1971). The microscopic appearance of waved-2 mouse hairs. *Genet. Res.* **17**, 257–60.

McLaren, A. (1972*a*). Germ cell differentiation in artificial chimaeras of mice. In *The Genetics of the Spermatozoon*, ed. R. A. Beatty & S. Gluecksohn-Waelsch, pp. 313–324. Edinburgh & New York.

McLaren, A. (1972*b*). Numerology of development. *Nature, Lond.* **239**, 274–6.

McLaren, A. (1972*c*). Late-labelling as an aid to chromosomal sexing of cultured mouse blood cells. *Cytogenetics* **11**, 35–45.

McLaren, A. (1975*a*). Sex chimaerism and germ cell distribution in a series of chimaeric mice. *J. Embryol. exp. Morph.* **33**, 205–16.

McLaren, A. (1975*b*). The independence of germ-cell genotype from somatic influence in chimaeric mice. *Genet. Res.* **25**, 83–7.

McLaren, A. & Biggers, J. D. (1958). Successful development and birth of mice cultivated *in vitro* as early embryos. *Nature, Lond.* **182**, 877–8.

McLaren, A. & Bowman, P. (1969). Mouse chimaeras derived from fusion of embryos differing by nine genetic factors. *Nature, Lond.* **224**, 238–40.

McLaren, A., Chandley, A. C. & Kofman-Alfaro, S. (1972). A study of meiotic germ cells in the gonads of foetal mouse chimaeras. *J. Embryol. exp. Morph.* **27**, 515–24.

McLaren, A., Gauld, I. K. & Bowman, P. (1973). A comparison between mice chimaeric and heterozygous for the X-linked gene *tabby*. *Nature, Lond.* **241**, 180–183.

McLaren, A. & Michie, D. (1956). Studies on the transfer of fertilized mouse eggs to uterine foster-mothers. I. Factors affecting the implantation and survival of native and transferred eggs. *J. exp. Biol.* **33**, 394–416.

Meo, T., Matsunaga, T. & Rijnbeek, A. M. (1973). On the mechanism of self-tolerance in embryo-fusion chimeras. *Transplant. Proc.* **5**, 1607–10.

Micklem, H. S., Ford, C. E., Evans, E. P. & Gray, J. (1966). Interrelationships of myeloid and lymphoid cells: studies with chromosome-marked cells transfused into lethally irradiated mice. *Proc. Roy. Soc. Lond. B.*, **165**, 78–102.

Micklem, H. S. & Loutit, J. F. (1966). *Tissue Grafting and Radiation*. New York & London: Academic Press.

Milet, R. G., Mukherjee, B. B. & Whitten, W. K. (1972). Cellular distribution and possible mechanism of sex-differentiation in XX/XY chimeric mice. *Can. J. Genet. Cytol.* **14**, 933–41.

Mintz, B. (1962*a*). Formation of genotypically mosaic mouse embryos. *Amer. Zool.* **2**, 432 (Abstr. 310).

Mintz, B. (1962*b*). Experimental recombination of cells in the developing mouse egg: normal and lethal mutant genotypes. *Amer. Zool.* **2**, 541–2. (Abstr.)

Mintz, B. (1962*c*). Experimental study of the developing mammalian egg: removal of the zona pellucida. *Science*, **138**, 594–5.

Mintz, B. (1964*a*). Formation of genetically mosaic mouse embryos, and early development of 'lethal (t^{12}/t^{12})–normal' mosaics. *J. exp. Zool.* **157**, 273–92.

Mintz, B. (1964*b*). Synthetic processes and early development in the mammalian egg. *J. exp. Zool.* **157**, 85–100.

Mintz, B. (1965*a*). Experimental genetic mosaicism in the mouse. In *Preimplantation Stages of Pregnancy*, ed. G. E. W. Wolstenholme & M. O'Connor, pp. 194–207. London: Churchill.

Mintz, B. (1965*b*). Genetic mosaicism in adult mice of quadriparental lineage. *Science* **148**, 1232–3.

Mintz, B. (1967*a*). Gene control of mammalian pigmentary differentiation. I. Clonal origin of melanocytes. *Proc. Nat. Acad. Sci. USA* **58**, 344–51.

Mintz, B. (1967*b*). Mammalian embryo culture. In *Methods in Developmental Biology*, ed. F. H. Wilt & N.K. Wessells, pp. 379–400. New York: Crowell.

Mintz, B. (1968). Hermaphroditism, sex chromosomal mosaicism and germ cell selection in allophenic mice. *J. Anim. Sci.* (Suppl. 1) **27**, 51–60.

Mintz, B. (1969*a*). Developmental mechanisms found in allophenic mice with sex chromosomal and pigmentary mosaicism. Birth defects: Original Article Series **5**, 11–22. In *First Conference on the Clinical Delineation of Birth Defects*, ed. D. Bergsma & V. McKusick. New York: National Foundation.

Mintz, B. (1969*b*). Gene control of the mouse pigmentary system. *Genetics*, **61**, (Suppl.) 41. (Abstr.)

Mintz, B. (1970*a*). Do cells fuse *in vivo*? In *In Vitro, Advances in Tissue Culture*, ed. C. Waymouth, **5**, 40–7. Baltimore: Williams & Wilkins.

Mintz, B. (1970*b*). Gene expression in allophenic mice. In *Control Mechanisms in the Expression of Cellular Phenotypes*, ed. H. A. Padykula, pp. 15–42. New York & London: Academic Press.

Mintz, B. (1970*c*). Gene control of differentiation of the mouse melanocyte system. *J. invest. Dermatol.* **54**, 93. (Abstr.)

Mintz, B. (1970*d*). Allophenic mice as test animals to detect tissue-specific histo-compatibility alloantigens or F₁ hybrid antigens. *Transplantation* **9**, 523–4.

Mintz, B. (1970*e*). Neoplasia and gene activity in allophenic mice. In *Genetic Concepts and Neoplasia*. Annual Symposium on Fundamental Cancer Research, pp. 477–517. Baltimore: Williams & Wilkins.

Mintz, B. (1971*a*). Allophenic mice of multi-embryo origin. In *Methods in Mammalian Embryology*, ed. J. C. Daniel Jr, pp. 186–214. San Francisco: Freeman.

Mintz, B. (1971*b*). The clonal basis of mammalian differentiation. In *Control Mechanisms of Growth and Differentiation*, ed. D. D. Davies & M. Balls, *Symposia of the Society for Experimental Biology*, **25**, pp. 345–370. London: Cambridge University Press.

Mintz, B. (1971*c*). Genetic mosaicism *in vivo*: development and disease in allophenic mice. *Fed. Proc.* **30**, 935–43.

Mintz, B. (1972*a*). Cellular expression of genes controlling susceptibility to neoplasia in mice. In *Cell Differentiation*, ed. R. R. Harris, P. Allin & D. Viza, pp. 176–181. Copenhagen: Munksgaard.

Mintz, B. (1972*b*). Clonal differentiation in early mammalian development. In *Molecular Genetics and Developmental Biology*, ed. M. Sussman, pp. 455–74. New Jersey: Prentice-Hall.

Mintz, B. (1972*c*). Clonal units of gene control in mammalian differentiation. In *Cell*

Differentiation, ed. R. R. Harris, P. Allin & D. Viza, pp. 267–71. Copenhagen: Munksgaard.

Mintz, B. & Baker, W. W. (1967). Normal mammalian muscle differentiation and gene control of isocitrate dehydrogenase synthesis. *Proc. Nat. Acad. Sci. USA*, **58**, 592–8.

Mintz, B., Custer, R. P. & Donnelly, A. J. (1971). Genetic diseases and developmental defects analyzed in allophenic mice. *Int. Rev. exp. Pathol.* **10**, 143–79.

Mintz, B., Domon, M., Hungerford, D. A. & Morrow, J. (1972). Seminal vesicle formation and specific male protein secretion by female cells in allophenic mice. *Science*, **175**, 657–9.

Mintz, B., Gearhart, J. D. & Guymont, A. O. (1973). Phytohemagglutinin-mediated blastomere aggregation and development of allophenic mice. *Dev. Biol.* **31**, 195–9.

Mintz, B. & Palm, J. (1965). Erythrocyte mosaicism and immunological tolerance in mice from aggregated eggs. *J. Cell Biol.* **27**, 66–7. (Abstr.)

Mintz, B. & Palm, J. (1969). Gene control of hematopoiesis. I. Erythrocyte mosaicism and permanent immunological tolerance in allophenic mice. *J. exp. Med.* **129**, 1013–27.

Mintz, B. & Sanyal, S. (1970). Clonal origin of the mouse visual retina mapped from genetically mosaic eyes. *Genetics* **64** (Suppl.), 43–4.

Mintz, B. & Silvers, W. K. (1967). 'Intrinsic' immunological tolerance in allophenic mice. *Science*, **158**, 1484–7.

Mintz, B. & Silvers, W. K. (1970). Histocompatibility antigens on melanoblasts and hair follicle cells. *Transplantation* **9**, 497–505.

Mintz, B. & Slemmer, G. (1969). Gene control of neoplasia. I. Genotypic mosaicism in normal and preneoplastic mammary glands of allophenic mice. *J. Nat. Cancer Inst.* **43**, 87–95.

Moore, W. J. & Mintz, B. (1972). Clonal model of vertebral column and skull development derived from genetically mosaic skeletons in allophenic mice. *Dev. Biol.* **27**, 55–70.

Moores, P. P. (1966). An Asiatic blood group chimera. Paper read at Blood Trans. Congr., Port Elizabeth. Cited by Race & Sanger (1968).

Morgan, T. H. (1927). The fusion of two eggs to produce one embryo. In *Experimental Embryology*, Chap. 20. New York: Columbia University Press.

Moustafa, L. A. (1974). Chimaeric rabbits from embryonic cell transplantation (38371). *Proc. Soc. exp. Biol. Med.* **147**, 485–8.

Moustafa, L. A. & Brinster, R. L. (1972*a*). The fate of transplanted cells in mouse blastocysts *in vitro*. *J. exp. Zool.* **181**, 181–92.

Moustafa, L. A. & Brinster, R. L. (1972*b*). Induced chimerism by transplanting embryonic cells into mouse blastocysts. *J. exp. Zool.* **181**, 193–202.

Mukherjee, B. B. & Milet, R. G. (1972). Nonrandom X-chromosome inactivation – an artifact of cell selection. *Proc. Nat. Acad. Sci. USA*, **69**, 37–9.

Mullen, R. J. & Carter, S. C. (1973). Efficiency of transplanting normal, zona-free and chimeric embryos to one and both uterine horns of inbred and hybrid mice. *Biol. Reprod.* **9**, 111–15.

Mullen, R. J. & Whitten, W. K. (1971). Relationship of genotype and degree of chimerism in coat color to sex ratios and gametogenesis in chimeric mice. *J. exp. Zool.* **178**, 165–76.

Mulnard, J. G. (1967). Analyse microcinématographique du développement de l'oeuf de souris du stade II au blastocyste. *Archs. Biol.*, *Liège*, **78**, 107–38.

Mulnard, J. G. (1971). Manipulation of cleaving mammalian embryos with special reference to a time-lapse cinematographic analysis of centrifuged and fused mouse eggs. *Adv. Biosci.* **6**, 255–74.

Mulnard, J. G. (1973). Formation de blastocystes chimériques par fusion d'embryons de Rat et de Souris au stade VIII. *C. r. hebd. Séanc. Acad. Sci., Paris*, **276**, 379–81.

Munro, A. J., Day, K. & Gardner, R. L. (1974). The use of chimaeric mice to search for gene transfer in the immune response. *Immunology* **27**, 525–30.

Myhre, B. A., Meyer, T., Opitz, J. M., Race, R. R., Sanger, R. & Greenwalt, T. J. (1965). Two populations of erythrocytes associated with XX/XY mosaicism. *Transfusion* (Philadelphia) **5**, 501–5.

Mystkowska, E. T. (1975). Development of mouse–bank vole interspecific chimaeric embryos. *J. Embryol. exp. Morph.*, **33**, 731–44.

Mystkowska, E. T. & Tarkowski, A. K. (1968). Observations on CBA-p/CBA-T6T6 mouse chimaeras. *J. Embryol. exp. Morph.* **20**, 33–52.

Mystkowska, E. T. & Tarkowski, A. K. (1970). Behaviour of germ cells and sexual differentiation in late embryonic and early postnatal mouse chimaeras. *J. Embryol. exp. Morph.* **23**, 395–405.

Nes, N. (1966). Diploid–triploid chimerism in a true hermaphrodite mink (*Mustela vison*). *Hereditas* **56**, 159–70.

Nesbitt, M. N. (1971). X-chromosome inactivation mosaicism in the mouse. *Dev. Biol.* **26**, 252–63.

Nesbitt, M. N. (1974). Chimeras *vs.* X inactivation mosaics: significance of differences in pigment distribution. *Dev. Biol.* **38**, 202–7.

Nesbitt, M. N. & Gartler, S. M. (1971). The applications of genetic mosaicism to developmental problems. *Ann. Rev. Genet.* **5**, 143–62.

Nieuwkoop, P. D. (1946). Experimental investigations on the origin and determination of the germ cells, and on the development of the lateral plates and germ ridges in urodeles. *Archs. néerl. Zool.* **8**, 1–205.

Ohno, S., Trujillo, J. M., Stenius, C., Christian, L. C. & Teplitz, R. L. (1962). Possible germ cell chimeras among newborn dizygotic twin calves (*Bos taurus*). *Cytogenetics* **1**, 258–65.

Okada, M., Kleinman, I. A. & Schneiderman, H. A. (1974). Chimeric *Drosophila* adults produced by transplantation of nuclei into specific regions of fertilized egg. *Dev. Biol.* **39**, 286–94.

Okada, T. S. (1955). Experimental studies on the differentiation of the endodermal organs in amphibia. IV. The differentiation of the intestine from the foregut. *Annotnes zool. jap.* **28**, 210–14.

Overzier, C. (1964). Ein XX/XY Hermaphrodit mit einem 'intratubulären Ei' und einem Gonadoblastom (gonocytom III). *Klin. Wochenschr.* **42**, 1052–60.

Padeh, B., Wysoki, M., Ayalon, N. & Soller, M. (1965). An XX/XY hermaphrodite in the goat. *Israel J. Med. Sci.* **1**, 1008–12.

Papp, von Z., Gardo, S., Herpay, G. & Arvay, A. (1970). Echter Hermaphroditismus mit Chromosomenmosaik 46,XX/46,XY. *Zbl. Gynäk.* **92**, 1183–9.

Park, I. & Jones, H. W. (1970). The phenotypic expression of 46,XX/46,XY genetic chimera in humans. *Internat. J. Gynec. Obstet.* **8**, 147. (Abstr. 107.)

Peterson, A. C. (1974). Chimaera mouse study shows absence of disease in genetically dystrophic muscle. *Nature, Lond.* **248**, 561–4.

Phillips, S. M., Martin, W. J., Shaw, A. R. & Wegmann, T. G. (1971). Serum-mediated immunological non-reactivity between histo-incompatible cells in tetraparental mice. *Nature, Lond.*, **234**, 146–8.

Phillips, S. M. & Wegmann, T. G. (1973). Active suppression as a possible mechanism of tolerance in tetraparental mice. *J. exp. Med.* **137**, 291–300.

Pighills, E., Hancock, J. L. & Hall, J. G. (1968). Attempted induction of chimaerism in sheep. *J. Reprod. Fert.* **17**, 543–7.

Polani, P. E. (1970). Hormonal and clinical aspects of hermaphroditism and the testicular feminizing syndrome in man. *Phil. Trans. Roy. Soc. Lond.* B, **259**, 187–204.

Race, R. R. & Sanger, R. (1968). *Blood Groups in Man*, 5th edition. Oxford & Edinburgh: Blackwell.

Ramseier, H. (1973). Immunization against abolition of transplantation tolerance. *Eur. J. Immunol.* **3**, 156–64.

Ransom, R., Hill, W. G. & Kacser, H. (1975). Cells, clones and patches: an analysis. In preparation.

Russell, A. S., Liburd, E. M. & Diener, E. (1974). *In vitro* suppression of cell mediated autoimmunity in NZB mice. *Nature, Lond.*, **249**, 43–5.

Russell, L. B. & Woodiel, F. N. (1966). A spontaneous mouse chimera formed from separate fertilization of two meiotic products of oogenesis. *Cytogenetics* **5**, 106–19.

Schmid, W. & Vischer, D. (1967). A malformed boy with double aneuploidy and diploid–triploid mosaicism 48,XXYY/71,XXXYY. *Cytogenetics* **6**, 145–55.

Segni, G. & Grossi-Bianchi, M. L. (1965). Un raro caso di ermafroditismo vero di tipo alterno con cromosomi XX/XY. *Minerva Pediat.* **17**, 983–9.

Shanfield, I., Young, R. B. & Hume, D. M. (1973). True hermaphroditism with XX/XY mosaicism. *J. Pediat.* **83**, 471–3.

Short, R. V. (1970). The bovine freemartin: a new look at an old problem. *Phil. Trans. Roy. Soc. Lond.* B, **259**, 141–7.

Short, R. V. (1972). Germ cell sex. In *The Genetics of the Spermatozoon*, ed. R. A. Beatty & S. Gluecksohn-Waelsch, pp. 325–345. Edinburgh & New York.

Silvers, W. K. & Russell, E. S. (1955). An experimental approach to action of genes at the agouti locus in the mouse. *J. exp. Zool.* **130**, 199–220.

Spemann, H. (1938). *Embryonic Development and Induction*. New Haven: Yale University Press.

Stern, C. (1968). Genetic mosaics in animals and Man. In *Genetic Mosaics and Other Essays*. Cambridge, Mass: Harvard University Press.

Stern, M. S. (1972). Experimental studies on the organization of the preimplantation mouse embryo. II. Reaggregation of disaggregated embryos. *J. Embryol. exp. Morph.* **28**, 255–61.

Stern, M. S. (1973). Chimaeras obtained by aggregation of mouse eggs with rat eggs. *Nature, Lond.* **243**, 472–3.

Stern, M. S. & Wilson, I. B. (1972). Experimental studies on the organization of the preimplantation mouse embryo. I. Fusion of asynchronously cleaving eggs. *J. Embryol. exp. Morph.* **28**, 247–54.

Subak-Sharpe, H., Bürk, R. R. & Pitts, J. D. (1969). Metabolic cooperation between biochemically marked mammalian cells in tissue culture. *J. Cell Sci.* **4**, 353–67.

Takagi, N. & Sasaki, M. (1975). Preferential inactivation of the paternally derived X chromosome in the extraembryonic membranes of the mouse. *Nature, Lond.*, **256**, 640–2.

Tarkowski, A. K. (1961). Mouse chimaeras developed from fused eggs. *Nature, Lond.*, **190**, 857–60.

Tarkowski, A. K. (1962). Inter-specific transfer of eggs between rat and mouse. *J. Embryol. exp. Morph.* **10**, 476–95.

Tarkowski, A. K. (1963). Studies on mouse chimaeras developed from eggs fused *in vitro*. *Natl. Cancer Inst. Monograph* **11**, 51–71.

Tarkowski, A. K. (1964a). True hermaphroditism in chimaeric mice. *J. Embryol. exp. Morph.* **12**, 735–57.

Tarkowski, A. K. (1964b). Patterns of pigmentation in experimentally produced mouse chimaeras. *J. Embryol. exp. Morph.* **12**, 575–85.

Tarkowski, A. K. (1965). Embryonic and postnatal development of mouse chimaeras. In *Preimplantation Stages of Pregnancy*, ed. G. E. W. Wolstenholme & M. O'Connor, pp. 183–93. London: Churchill.

Tarkowski, A. K. (1969). Consequences of sex chromosome chimerism for sexual differentiation in mammals. *Annls. Embryol. Morph.*, Suppl. **1**, 211–22.

Tarkowski, A. K. (1970a). Are the genetic factors controlling sexual differentiation of somatic and germinal tissues of a mammalian gonad stable or labile? *Fogarty Int. Cent. Proc.* **2**. US Govt Printing Office.

Tarkowski, A. K. (1970b). Germ cells in natural and experimental chimaeras in mammals. *Phil. Trans. R. Soc. Lond.* B, **259**, 107–11.

Tarkowski, A. K. & Wroblewska, J. (1967). Development of blastomeres of mouse eggs isolated at the 4- and 8-cell stage. *J. Embryol. exp. Morph.* **18**, 155–80.

Tettenborn, U., Dofuku, R. & Ohno, S. (1971). Noninducible phenotype exhibited by a proportion of female mice heterozygous for the X-linked testicular feminization mutation. *Nature, Lond.* **234**, 37–40.

Tucker, E. M., Moor, R. M. & Rowson, L. E. A. (1974). Tetraparental sheep chimaeras induced by blastomere transplantation. Changes in blood type with age. *Immunology* **26**, 613–21.

Tuffrey, M. & Barnes, R. D. (1972). Manipulation of fertilised mouse eggs applied to immunology. In *Advances in Experimental Medicine and Biology*, ed. B. D. Jankovic, **29**, pp. 637–41. New York & London: Plenum Press.

Tuffrey, M. & Barnes, R. D. (1973). Investigation of autoimmune control in tetraparental chimaeric mice. In *Immunology of Reproduction*, pp. 544–51. Sofia: Bulgarian Academy of Sciences.

Tuffrey, M., Barnes, R. D., Evans, E. P. & Ford, C. E. (1973a). Dominance of AKR lymphocytes in tetraparental AKR↔CBA-T6T6 chimaeras. *Nature New Biol.*, **243**, 207–8.

Tuffrey, M., Holliday, J. & Barnes, R. D. (1974). Renal lesions in tetraparental mouse chimaeras stimulated in NZB by injection of CFW spleen cells. *J. Pathol.* **113**, 61–7.

Tuffrey, M., Kingman, J. & Barnes, R. D. (1973*b*). Disease in the progeny of ovum fusion derived tetraparental NZB↔CFW chimaeras. *J. Pathol.* **111**, 213–19.

Vickers, A. D. (1967). A direct measurement of the sex-ratio in mouse blastocysts. *J. Reprod. Fert.* **13**, 375–6.

Vigier, B., Prepin, J. & Jost, A. (1972). Absence de corrélation entre le chimérisme XX/XY dans le foie et les premiers signes du freemartinisme chez le foetus de veau. *Cytogenetics* **11**, 81–101.

Wegmann, T. G. (1970). Enzyme patterns in tetraparental mouse liver. *Nature, Lond.*, **225**, 462–3.

Wegmann, T. G. & Gilman, J. G. (1970). Chimaerism for three genetic systems in tetraparental mice. *Dev. Biol.* **21**, 281–91.

Wegmann, T. G., Hellström, I. & Hellström, K. E. (1971*a*). Immunological tolerance: 'Forbidden clones' allowed in tetraparental mice. *Proc. Nat. Acad. Sci. USA*, **68**, 1644–7.

Wegmann, T. G., LaVail, M. M. & Sidman, R. L. (1971*b*). Patchy retinal degeneration in tetraparental mice. *Nature, Lond.*, **230**, 333–4.

West, J. D. (1975). A theoretical approach to the relation between patch size and clone size in chimaeric tissue. *J. theor. Biol.* **50**, 153–60

West, J. D. (1976*a*). Clonal development of the retinal epithelium of mouse chimaeras and X-inactivation mosaics. *J. Embryol. exp. Morph.* in press.

West, J. D. (1976*b*). Red-blood-cell selection in chimaeric mice. *Exp. Hematol.* in press.

West, J. D. (1976*c*). Patches in the livers of chimaeric mice. *J. Embryol. exp. Morph.* in press.

West, J. D. (1976*d*). Distortion of patches of retinal degeneration in chimaeric mice. *J. Embryol. exp. Morph.* in press.

West, J. D. & McLaren, A. (1976). The distribution of melanocytes in the dorsal coats of a series of chimaeric mice. *J. Embryol. exp. Morph.* **35,** 87-93.

Whitten, W. K. (1956). Culture of tubal mouse ova. *Nature, Lond.*, **177**, 96.

Whitten, W. K. (1957). Culture of tubal ova. *Nature, Lond.*, **179**, 1081–2.

Wills, E. A., Tuffrey, M. A. & Barnes, R. D. (1975). C-type murine leukaemia virus particles in tetraparental AKR↔CBA chimaeras. *Clin. exp. Immunol.* **20**, 563–9.

Wilson, I. B., Bolton, E. & Cuttler, R. H. (1972). Preimplantation differentiation in the mouse egg as revealed by microinjection of vital markers. *J. Embryol. exp. Morph.*, **27**, 467–79.

Wolff, G. L. (1971). Genetic modification of homeostatic regulation in the mouse, *Amer. Nat.* **105**, 241–52.

Wolpert, L. & Gingell, D. (1970). Striping and the pattern of melanocyte cells in chimaeric mice. *J. theor. Biol.* **29**, 147–50.

Zeilmaker, G. (1973). Fusion of rat and mouse morulae and formation of chimaeric blastocysts. *Nature, Lond.*, **242**, 115–16.

Zuelzer, W. W., Beattie, K. M. & Reisman, L. E. (1964). Generalized unbalanced mosaicism attributable to dispermy and probable fertilization of a polar body. *Am. J. Hum. Genet.* **16**, 38–51.

Author Index

Where sequences of page numbers are separated only by spaces rather than commas, this indicates that on all these pages the author is a co-author of the reference given in brackets at the end of the sequences.

Subject Index